A New Family of CMOS Cascode-Free Amplifiers with High Energy-Efficiency and Improved Gain

Ricardo Filipe Sereno Póvoa
João Carlos da Palma Goes
Nuno Cavaco Gomes Horta

A New Family of CMOS Cascode-Free Amplifiers with High Energy-Efficiency and Improved Gain

 Springer

Ricardo Filipe Sereno Póvoa
Instituto de Telecomunicações
Instituto Superior Técnico
Universidade de Lisboa
Lisboa, Portugal

João Carlos da Palma Goes
UNINOVA, Faculdade de Ciências e
Tecnologia, Universidade Nova de Lisboa
Lisboa, Portugal

Nuno Cavaco Gomes Horta
Instituto de Telecomunicações
Instituto Superior Técnico
Universidade de Lisboa
Lisboa, Portugal

ISBN 978-3-030-06992-6   ISBN 978-3-319-95207-9   (eBook)
https://doi.org/10.1007/978-3-319-95207-9

This Springer imprint is published by the registered company Springer Nature Switzerland AG
The registered company address is: Gewerbestrasse 11, 6330 Cham, Switzerland

Ricardo Póvoa
To my dear Marta, for everything that is worth fighting for.

João Goes
To Valentim and Madalena.

Nuno Horta
To Carla, João, and Tiago.

Preface

Amplifiers are important in several electronic systems and processing chains, e.g., radio-frequency transceivers in wireless networks, data acquisition channels, or analog-to-digital converters (ADCs). Therefore, the applications of amplifiers spread from analog to mixed-signal design. Desirably, amplifiers allow an efficient amplification of small-signals, without adding significant noise to the chain, and can also operate as signal comparators in ADCs. The power reduction necessity, the intrinsic gain reduction and high variability, with the low-supply voltage trend of modern CMOS technologies, has driven the evermore challenging design of amplifiers to implement multiple gain stages and possibly one output driver stage. Regarding single-stage amplifiers, commonly more power efficient, high gains are normally achieved using cascode devices or cascaded stages. However, this leads to reduced output swings (OS) due to the lower supply voltages in hand with the stacking of devices. In residue amplifiers, inside ADCs, one possible solution is dynamic amplification. Dynamic CMOS amplifiers are proposed by Copeland and Rabaey, in 1979, in which the idea is that the bias current is not constant but changed during amplification. This property is important, particularly in switched-capacitor circuits, where the amplifier can be biased in strong inversion with a large current in the beginning and then continuously reduced the current toward weak inversion, until the power is practically cut off, maximizing both the gain and the OS of the amplifier. This work addresses the need for energy-efficient amplifiers and gain enhancement strategies, compatible with lower supply voltages, by proposing a complete new family of single-stage cascode-free amplifiers, with design, optimization, and experimental evaluation. The energy efficiency and topological potential are maximized through advanced automation carried by AIDA, an analog IC design and optimization framework, based on computational intelligence. The topologies are proposed using the UMC 130 nm CMOS technology for proof of concept: voltage combiners (VC) biased operational transconductance amplifier (OTA), VC biased OTA with current starving for higher gain and energy efficiency, folded VC biased OTA for lower-voltage sources, and a dynamic VC biased OTA targeting ADCs, with gains above 50 dB and energy-efficient figure-of-merit values of 1024,

1102, 2279, and 1349 MHz $\times$ pF/mA, correspondingly. The presented results are beyond what is achievable with a classic folded-cascode amplifier, and, regarding dynamic amplifiers, the proposed solutions clearly contribute to advances in the state of the art. This work is organized in six chapters. Chapter 1 presents a brief introduction with the motivation and context to develop and propose new amplifier topologies, with high energy efficiency and gain improvement, particularly in the environment of the Internet of Things, emphasizing the wide field of applications of amplifiers. Chapter 2 discusses the background and the state of the art of single-stage amplifiers, providing the context of the developed work. The main topologies are detailed: the telescopic-cascode, the mirrored-cascode, and the folded-cascode amplifiers, followed by the recycling folded-cascode amplifier with the corresponding improvements and surveying the concept of dynamic amplification in CMOS technologies. The performance metrics are summarized, and a throughout comparison of the prior art is provided. Chapter 3 presents the architectures proposed in this work and shows circuit implementations in detail. The basic voltage-combiner structure is presented and detailed both at analytical and simulation levels. The complete set of amplifiers is shown, both in terms of topological description and analytic analysis. Sizing strategies and initial designs, i.e., first approaches that guarantee functional circuits, are presented with results at simulation level, validating the proposed topologies with applied noise modeling. Chapter 4 presents the optimization framework, AIDA; the complete setup for the optimization, regarding the objectives, specifications, variables, and ranges; and post-optimization simulation results for selected sizing solutions, positioning the proposed circuits in the context of the state of the art of single-stage amplifiers. Chapter 5 presents the printed circuit board developed to properly measure the fabricated circuits during the course of the present dissertation. The prototyped solutions are detailed: layout, post-layout, Monte Carlo simulations, experimental measurements, and comparison with the state of the art. Finally, Chapter 6 draws the conclusions, compounded with a summary of all the achieved developments, positioning the proposed topologies relative to the state of the art of single-stage amplification architectures.

Lisboa, Portugal   Ricardo Filipe Sereno Póvoa
                                                                                        João Carlos da Palma Goes
                                                                                        Nuno Cavaco Gomes Horta

Acknowledgments

If one considers a short-term signal in frequency, one knows that it extends intensively in time. The same principle applies to the expression of our gratitude to the following wonderful human beings, to whom we dedicate this work.

Ricardo Póvoa would like to thank his family for all the long-term love and support; his closest friends, David Raposo, Inês Vitorino, João Doroana, João S. Braz, Luís Cruz, Luís Pinto, Marta Taveira, Patrícia Madeira, Rafael Cabral, Rafaela S. Braz, and Vitor Camacho; Prof. Doutor Nuno Cavaco Gomes Horta and Prof. Doutor João Carlos da Palma Goes, for all the confidence, enthusiasm, and guidance throughout the development of this work; his friends and colleges from Instituto Superior Técnico and Instituto de Telecomunicações, Nuno Lourenço, Ricardo Martins, António Canelas, Mauro Santos, Marco Pereira, and Mario Assunção; and the expert advices from Prof. Doutor João M. Vaz, Prof. Doutor Jorge M. Guilherme, Prof. Doutor Jorge M. Fernandes, and Prof. Doutor Marcelino B. Santos.

Finally, the authors would like to express their gratitude for the financial support that made this work possible. This work was supported in part by Fundação para a Ciência e Tecnologia (Grant FCT SFRH/BD/133662/2017) and by Instituto de Telecomunicações (Research Project RAPID UID/EEA/50008/2013).

Contents

List of Abbreviations

AA	Active area
AC	Alternate current
ADC	Analog-to-digital converter
AIDA	Analog Integrated Circuit Design Automation
AMG	Analog module generator
CD	Common-drain
CG	Common gate
CMFB	Common-mode feedback
CMOS	Complementary metal-oxide-semiconductor
CMRR	Common-mode rejection ratio
CNT	Carbon nanotube
COB	Chip-on-board
CPU	Central processing unit
CS	Common source
DC	Direct current
DEL	Saturation margin
DFT	Discrete Fourier transform
ESD	Electrostatic discharge
FCA	Folded-cascode amplifier
FF	Fast NMOS and fast PMOS
FFT	Fast Fourier transform
FNSP	Fast NMOS and slow PMOS
FOM	Figure of merit
GBW	Gain-bandwidth product
GDS	Geometric data stream
HVHT	High voltage high temperature
HVLT	High voltage low temperature
IC	Integrated circuits
IOT	Internet of Things

LVHT	Low voltage high temperature
LVLT	Low voltage low temperature
MC	Monte Carlo
MCA	Mirrored-cascode amplifier
MOSFET	Metal-oxide-semiconductor field-effect transistor
NMOS	n-channel MOSFET
NSGA	Non-sorting genetic algorithm
OPAMP	Operational amplifier
OPDK	Organic process design kit
OS	Output swing
OTA	Operational transconductance amplifier
OTFT	Organic thin-film transistor
OVD	Overdrive voltage
PCB	Printed circuit board
PM	Phase margin
PMOS	p-channel MOSFET
POF	Pareto optimal front
PSRR	Power-supply rejection ratio
PVT	Process, voltage, and temperature
RAM	Random-access memory
RF	Radio-frequency
RFCA	Recycling folded-cascode amplifier
RFID	Radio-frequency identification
RMS	Root mean square
SAR	Successive approximation register
SC	Switched-capacitor
SMA	Surface Mount Adhesive
SMD	Surface Mount Device
SNFP	Slow NMOS and fast PMOS
SOIC	Small Outline Integrated Circuit
SR	Slew rate
SS	Slow NMOS and slow PMOS
S/H	Sample-and-hold
TCA	Telescopic-cascode amplifier
TF	Transfer function
VC	Voltage combiner
VOS	Offset voltage
VT	Threshold voltage
UMC	United Microelectronics Corporation
WF	Width-*per*-finger

List of Symbols

a_v	Voltage gain
a_{cm}	Common-mode voltage gain
a_{dm}	Differential-mode voltage gain
b	Mirroring factor
c	Light speed in vacuum
c_{db}	Small-signal drain-bulk capacitance
c_{gd}	Small-signal gate-drain capacitance
c_{gs}	Small-signal gate-source capacitance
c_l	Load capacitance
c_{ox}	Gate oxide capacitance *per* unit of area
c_s	Sampling capacitance
cmfb	Common-mode feedback
f	Frequency
$f_{-3\,db}$	Bandwidth
f_c	Corner frequency
f_m	Objective function
g_m	Constraints function
gds	Small-signal drain-source conductance
gm	Small-signal transconductance
i	Electric current
i_d	Drain current
i_{dd}	Current consumption
i_{dsat}	Saturation current
i_o	Output current
i_{ref}	Reference current
k	Boltzmann constant
l	Channel length
nf	Channel number of fingers
p_m	Performance figures
q	Electric charge

r_{eq}	Equivalent resistance
r_o	Output resistance
t	Time unit
t_{ox}	Gate oxide thickness
v_{bias}	Bias voltage
v_{cm}	Common-mode voltage
v_d	Differential-mode voltage
v_{dd}	Positive supply voltage
v_{ds}	Drain-source voltage
vds_{sat}	Saturation voltage
v_{gs}	Gate-source voltage
v_i	Input voltage
v_o	Output voltage
v_{reg}	Voltage regulator output voltage
v_{ss}	Negative supply voltage
v_t	Threshold voltage
w	Channel width
y	Admittance
ε	Epsilon
ε_{ox}	Oxide permittivity
κ	Velocity-saturation degree
λ	Wavelength
μ	Channel mobility
ξ_c	Velocity-saturation threshold
σ	Sigma
ω_p	Pole frequency

Chapter 1
Introduction

The motivation to this work relies on nowadays trend to incorporate complete systems in portable and highly autonomous electronic equipment, i.e., longer-lasting battery-powered circuitry, to assist the everyday life of populations. Indeed, the need for low-power circuits is directly related with the energy efficiency of the amplifiers, used in the different building blocks employed in an analog processing chain of a given circuit or electronic system. In fact, amplifiers are fundamental components, not only in analog but also in digital and mixed-signal processing.

The need of power reduction in electronic circuitry, visible in both professional and personal applications of the everyday life, together with the tendency of modern complementary metal-oxide-semiconductor (CMOS) technologies to use low-supply voltages, drives the evermore challenging design of amplifiers [1]. The design of amplifiers, in general, follows the straightforward concept of multiple cascaded gain stages, since the native cascode approach suffers from output swing reduction due to transistor stacking. Multistage amplification architectures are commonly terminated with the implementation of output driver stages, increasing the complexity of the circuits and, therefore, their mass production cost. Reducing power in analog circuitry is mainly related with the energy efficiency of the amplifiers. Hence, they are fundamental and key components, not only in analog but also in digital and mixed-signal processing chains, as shown in Figs. 1.1 and 1.2 [2–5]. Therefore, the need for high performance together with high energy efficiency urges in the particular design of amplifiers. As a consequence, new and improved topologies have to be proposed.

Particularly, in the novel paradigm of the Internet of Things (IOT) that is rapidly gaining significance in the scenario of modern wireless telecommunications, analog amplifiers play a most significant role, and specifically, the energy efficiency is of major importance [6–8]. The IOT is described in the following paragraphs, and the importance of high-energy-efficient amplifiers, particularly in this context, is clarified. The basic idea of the IOT is the pervasive presence around us of a variety of things or objects, e.g., radio-frequency identification (RFID) tags, sensors, actuators,

© Springer International Publishing AG, part of Springer Nature 2019
R. F. S. Póvoa et al., *A New Family of CMOS Cascode-Free Amplifiers with High Energy-Efficiency and Improved Gain*,
https://doi.org/10.1007/978-3-319-95207-9_1

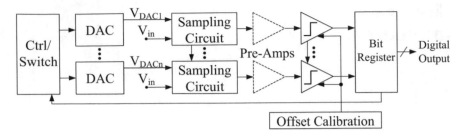

Fig. 1.1 Applications of amplifiers in modern electronics: architecture of successive approximation register (SAR) analog-to-digital converter (ADC) [2]

Fig. 1.2 Applications of
amplifiers (block G) in
modern electronics: pipeline
ADC diagram block [3]

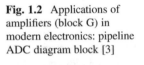

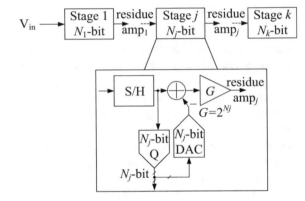

or mobile phones, which, through unique addressing schemes, are able to interact
with each other and cooperate with their neighbors to reach common goals [7]. In
other words, each electronic device operates within the Internet with a dedicated
identification, through radio, i.e., RFID, and all the connected devices are identified
in a common database, looking toward an easier and effortless everyday life.
Indisputably, the main strength of the IOT idea is the high impact it has on several
aspects of everyday life and on the behavior of potential users. Moreover, the
detection of the physical status of things through sensors, together with collection
and processing of detailed data, allows an immediate response to changes in the real
world. This fully interactive and responsive network yields immense potential for
citizens, consumers, and business [9]. The real world becomes more accessible
through computers and networked devices in business as well as everyday scenarios,
as Fig. 1.3 illustrates [10].

Specifically, from the point of view of the private user, the most obvious effects of
the IOT introduction are visible in both working and domestic fields. In this context,
domestic electronics, i.e., domotics, home automation, assisted living, e-health, and
enhanced learning, are only a few examples of possible application scenarios, in
which the new paradigm is playing a leading role in the present time. Similarly, from
the perspective of business users, the most apparent consequences are equally visible
in fields such as automation and industrial manufacturing, logistics, business and

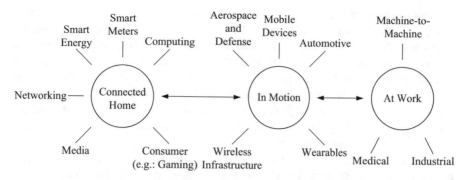

Fig. 1.3 Field of applications for the IOT in the everyday life

Fig. 1.4 Context and applications for the IOT

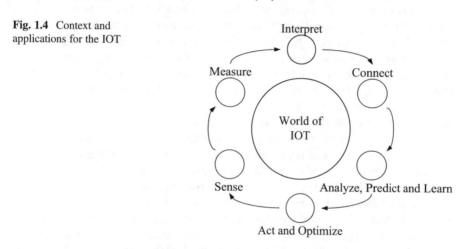

process management, as well as intelligent transportation of people and goods [6]. The overall functioning of the infrastructure that supports nowadays IOT, as proposed in [11], contemplates a finite set of stages that represents, in practical manners, specific fields of electronics engineering that, when compounded together, allow the system to operate with demanding robustness and fidelity. The representation of this model, shown in Fig. 1.4, possesses six stages that follow a recursive flow, but do not necessarily follow the depicted order, i.e., sensing, measuring, interpreting, connectivity, analysis, action, and optimization. In order to support the variety and amount of computational and communicational effort, the IOT is placing new demands on this network infrastructure itself, therefore, in each of the stages depicted in Fig. 1.4. Applications require high-speed connections, an extremely low latency response, and integrated security, while organizations need primarily a flexible and scalable wired and wireless network infrastructure [12].

In this context, the need for amplifiers with high energy efficiency becomes evident, i.e., a preponderant part of applications requires wireless, remote, or mobile solutions; thus, battery life can have significant implications on the overall cost of production and, consequently, in the cost of ownership. In detail, electronic amplifiers play their most important role in measurement equipment, connectivity solutions, and sensor interfaces [13, 14]. In parallel with the need for energy efficiency, the concern over gain enhancement strategies is of utmost relevance, due to the fact that designs targeting deeper nanoscale technologies, particularly beyond the 65 nm node, suffer a reduction of the transistors intrinsic gain, i.e., gm/gds where gm and gds represent, correspondingly, the main transconductance and the output conductance of the devices, to a value below 10 V/V, i.e., 20 dB, and the intrinsic gain variability rises about 10 dB [15, 16]. This concern is present in single-stage and in multistage amplifiers.

This work addresses these two important matters, i.e., the need for energy efficiency and the development of gain enhancement strategies free from cascode implementations, following the tendency of lower-voltage supplies, by proposing a complete new family of single-stage amplifier architectures to be implemented as a stand-alone or as part of multistage amplifier chains, embedded in high-performance ADCs, in radio-frequency (RF) demodulators and transceivers, as well as in the nowadays intensifying standpoint of IOT. In particular, this work proposes a complete new family of cascode-free single-stage amplifiers with high gain and high energy efficiency through the usage of voltage-combiner biasing structures, in replacement of the traditional current sources of single-stage amplifiers. This has a twofold effect: (a) additional gain is provided, and (b) the differential-pair devices act as a common source and a common gate simultaneously; this way the energy efficiency of the circuit is improved by means of increasing the gain-bandwidth product. This work comprehends static approaches for high and also for low biasing voltages, therefore exploring a set of different configurations. The concept is proved through the development and experimental evaluation of several integrated prototypes. Secondarily, a fully dynamic architecture biased by voltage combiners is also addressed, targeting low-power, moderate-resolution, and high-speed analog-to-digital converters, as well as high-performance switched-capacitor (SC) filters, in which results are presented at simulation level only. The experimental test and evaluation of a fully dynamic amplifier in a real- and continuous-time test environment are fairly difficult, i.e., in order to be carried out with satisfactory precision, the amplifier must be embedded in a physical implementation of an ADC; hence, it is considered out of the scope of this work. For proof of concept, the United Microelectronics Corporation (UMC) 130 nm technological node is selected and considered with standard threshold voltage (VT) devices only. The performance potential of the high-gain and energy-efficient designs is explored through the usage of advanced optimization techniques, implemented by analog integrated circuit design automation (AIDA). AIDA is an automatic analog integrated circuits (IC) design framework based on computational intelligence and electrical simulation and evaluation, developed to explore the design trade-offs involved in the design of analog electronic circuits [17]. The performance specifications targeted in this work are summarized as follows: the open-circuit gain regards values above 46 dB, proving the concept of gain enhancement through the usage of voltage combiners in replacement to the typical static approach to bias the differential pair, considering, as reference, the 40 dB open-circuit voltage gain of a single-stage symmetrical

CMOS OTA [18], for classically established technology nodes, e.g., the 130 nm technology. The extension of the proposed architectures to deeper nanoscale technology nodes, e.g., 65 or 40 nm, is relatively straightforward. The gain-bandwidth product and current consumption are the performance metrics to be improved in a way that translates into an energy-efficient figure of merit (FOM) higher than 1000 MHz × pF/mA, surpassing what is achievable with classic folded-cascode amplifier topologies, typically between 700 and 900 MHz × pF/mA, when properly optimized for FOM.

References

1. Online, 2017: https://www.imec-int.com/en/home.
2. H. Wei, et al., "An 8-b 400-MS/s 2-b-Per-Cycle SAR ADC With Resistive DAC," in IEEE Journal of Solid-State Circuits, Vol. 47, Issue 11, Page(s): 2535–2763, Nov. 2012. DOI: https://doi.org/10.1109/JSSC.2012.2214181.
3. M. Figueiredo, et al., "Reference-Free CMOS Pipeline Analog-to-Digital Converters," Springer, 2013. DOI: https://doi.org/10.1007/978-1-4614-3467-2-2.
4. V. Kopta, et al., "An Approximate Zero IF FM-UWB Receiver for High Density Wireless Sensor Networks," in IEEE Transactions on Microwave Theory and Techniques, Vol 65, Issue 2, Page(s): 374–385, Feb. 2017. DOI: https://doi.org/10.1109/TMTT.2016.2645568.
5. W. Gao, et al., "A Novel Front-End ASIC With Post Digital Filtering and Calibration for CZT-Based PET Detector," in Advancements in Nuclear Instrumentation Measurement Methods and their Applications (ANIMMA), Page(s): 1–5, Apr. 2015. DOI: https://doi.org/10.1109/ANIMMA.2015.7465547.
6. L. Atzori, et al., "The Internet of Things: A Survey," in Elsevier Computer Networks, Vol. 54, Issue 15, Page(s): 2787–2805, Oct. 2010. DOI: https://doi.org/10.1016/j.comnet.2010.05.010.
7. D. Giusto, et al., "The Internet of Things," Springer, 2010. DOI: https://doi.org/10.1007/978-1-4419-1674-7.
8. Online, 2017: http://www.analog.com/en/applications/markets/internet-of-things.html.
9. D. Uckelmann, et al., "An Architectural Approach towards the Future Internet of Things," Springer. Mar. 2011. DOI: https://doi.org/10.1007/978-3-642-19157-2_1.
10. Online, 2017: http://www.skyworksinc.com/ProductsLanding.aspx.
11. Online, 2017: http://www.analog.com/en/applications/technology.html.
12. Online, 2017: http://www.cisco.com/c/en/us/solutions/internet-of-things/iot-network-connectivity.html.
13. Online, 2017: http://www.analog.com/en/products/application-specific/medical/ecg/AD8233.html#product-over view.
14. Online, 2017: http://www.analog.com/media/en/technical-documentation/data-sheets/ADXL362.pdf.
15. J. Pekarik, et al., "RFCAMOS Technology from 0.25um to 65nm: The State of the Art," in IEEE Custom Integrated Circuits Conference, Page(s): 217–224, Oct. 2004. DOI: https://doi.org/10.1109/CICC.2004.1358782.
16. A. Baschirotto, et al., "Low Power Analog Design in Scaled Technologies", 2009.
17. N. Lourenço, et al., "AIDA: Robust Analog Circuit-Level Sizing and In-Circuit Layout Generation," in Integration, the VLSI Journal, 2016. DOI: https://doi.org/10.1016/j.vlsi.2016.04.009.
18. W. Sansen, "Analog Design Essentials," Springer, 2006. ISBN: 978-0-387-25747-1.

Chapter 2
Background and State of the Art

2.1 Background and Initial Considerations

Single-stage amplifiers are, in general, more power-efficient than multistage amplifiers and, in particular, two-stage amplifiers, if small and purely capacitive loads are considered, i.e., in the order of picofarads. In fact, single-stage amplifiers have gain-bandwidth product (GBW) values in the order of hundreds of megahertz, while multistage amplifiers present GBWs around tens of megahertz, for the same power consumption and load. Moreover, single-stage amplifiers present, in general, a higher ability to reject signals that are common to both inputs. For these principal reasons, this work is inserted in the context of single-stage amplifiers. Nevertheless, the proposed solutions can also be incorporated as first stages of multistage amplification architectures, without loss of impact and importance. Still in the context of single-stage amplifiers, cascode techniques can be used to achieve higher gains even when combined with other techniques, e.g., feed-forward [1], positive-feedback [2, 3], or regulated feed-forward [4]. The cascode approach, however, leads to lower output swings due to the increased number of transistors in stack, together with the tendency of reducing the supply voltages as exposed in [5]. Maintaining high gains but, simultaneously, requiring high-output swings leads to the natural choice of employing two-stage amplifiers applying cascode techniques only in the first stage, paying a higher price in terms of power consumption and active area. Indeed, the complexity and production cost of multistage amplifiers increase proportionally with the number of stages of the amplifier, i.e., active area, whereas single-stage amplifiers do not require compensation, which represents one less design variable and ultimately eases the design process. In addition, if proper pole compensation is not used in two-stage amplifiers, from 30% to 50% of power efficiency may be lost as it is clearly demonstrated in [6].

The present work addresses both the need for better energy efficiency and the necessity of improving the gain of single-stage amplifiers in a way that the output

© Springer International Publishing AG, part of Springer Nature 2019
R. F. S. Póvoa et al., *A New Family of CMOS Cascode-Free Amplifiers with High Energy-Efficiency and Improved Gain*,
https://doi.org/10.1007/978-3-319-95207-9_2

swing is not affected, as what happens in the case of cascode architectures, especially when low-supply voltages are employed, emphasized by the reduction of the transistors intrinsic gain, i.e., gm/gds, of deeper nanoscale technologies, through the proposal of a complete new family of single-stage amplifiers with high energy efficiency and gain enhancement.

2.2 Overview on Amplifiers

In the next paragraphs, the descriptions are focused on amplifiers. First, the operational amplifier (OpAmp) is described, followed by the analysis of basic single-stage and single-ended configurations. Regarding both differential and symmetric structures, the concept of "half circuit" is introduced, which usefully simplifies the analytical analysis of this type of circuits.

2.2.1 Operational Amplifiers

In theory, bipolar transistors offer many advantages in the design of OpAmps over CMOS technologies, e.g., a higher transconductance for a given current; a higher intrinsic gain, i.e., gm/r_o; higher speed, lower input-referred offset voltage; lower output impedance, i.e., assuming a conventional common-drain (class-A) output stage; and lower input-referred noise voltage. This way, OpAmps made from bipolar transistors offer, in many cases, the best performance. In fact, general purpose bipolar OpAmps, e.g., the 741 operational amplifier, became commercially important and are still used now, particularly in the academic context within the early years of development. However, CMOS technologies have become dominant, especially in the first two decades of the twenty-first century, in analog, digital, and mixed-signal electronics, due to the fact that CMOS circuits are smaller and dissipate considerably less power than their bipolar counterparts. With the growth of wireless and portable systems, and in the context of IOT, this fact is of utmost importance. Thus, to reduce system cost and increase portability, both analog and digital circuits are now often integrated together, providing a strong incentive to use CMOS OpAmps [7].

Taking into consideration an amplifier as a completely definable circuit, an ideal OpAmp is a circuit that has a differential input, infinite voltage gain, infinite input resistance, and zero output resistance, corresponding to an ideal voltage-controlled voltage source. The electrical scheme and representation of an ideal OpAmp is shown in Fig. 2.1. Generically, the gain considered in Fig. 2.1, A_v, refers only to the differential voltage applied to the inputs of the amplifier, hence defined as the differential gain. However, this is not enough, since the common-mode voltage, v_{cm}, defined as the constant value applied to both inputs at the same time, is also amplified

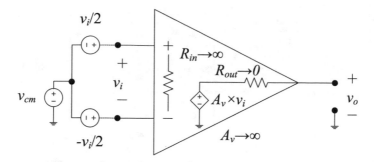

Fig. 2.1 Ideal OpAmp internal circuit

in practical realizations. The gain that applies to the common mode is defined as the common-mode gain and ideally is zero.

The differential input, v_d, is defined in (2.1), the common-mode voltage is defined in (2.2), and the output voltage is given by (2.3), considering the contributions of the differential gain, A_{dm}, and common-mode gain, A_{cm}. Following the nomenclature in Fig. 2.1, if a fully differential configuration is considered, the differential voltage gain is, in theory, twice the value of the single-ended configuration gain, as described further.

$$v_d = v^+ - v^- = v_i \tag{2.1}$$

$$v_{cm} = (v^+ + v^-)/2 \tag{2.2}$$

$$v_o = A_{dm}v_i + A_{cm}v_{cm} \tag{2.3}$$

2.2.2 Single-Stage Amplifiers

Virtually, every operational amplifier design comprises a single-stage amplification scheme inside the circuit diagram. Starting with a simple design, the common-source (CS) amplifier is the most popular gain stage, especially when high-input impedance is desired, since the input is applied to the gate of a metal-oxide-semiconductor field-effect transistor (MOSFET) [8].

A CS amplifier with resistive load, shown in Fig. 2.2, is the most basic amplifier topology that can be designed. The n-channel MOSFET (NMOS) transistor is fed by the current flowing through $R0$, while the latter sets the direct current (DC) output voltage. The small-signal equivalent circuit is shown in Fig. 2.2, and the gain of this configuration can be given by (2.4), where r_o represents the drain-source resistance of the transistor.

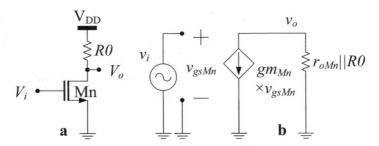

Fig. 2.2 CS amplifier with resistive load: (**a**) circuit electrical scheme; (**b**) small-signal simplified circuit

$$A_v = v_o/v_i = -\text{gm}_{\text{Mn}}(r_{o_{\text{Mn}}}\|R0) \tag{2.4}$$

The CS amplifier with resistive load, shown in Fig. 2.2, is fairly simple and understandable, yet this circuit is simply not practical in terms of implementation. The most important drawback, apart from the process variability of the resistor with direct impact on the performance of the circuit, is the fact that the gain is proportional to the area occupied by the resistor, which is fairly inconvenient. Moreover, the gain increases proportionally to the DC voltage drop in $R0$, i.e., if one wants to increase the gain, one has to increase the value of $R0$, i.e., the voltage drop between the terminals of the resistor, which is also inconvenient in the majority of the cases, e.g., due to output swing and when considering cascading structures.

Instead of a resistor, a transistor can be used if in diode configuration, i.e., with the gate connected to the drain. This is designated by diode-connected active load, in the sense that the load is a circuit component made up of active devices, as opposed to the resistor which is a passive device, hence designated by passive load. In fact, the active load is intended to present high small-signal impedance without requiring a large DC voltage drop. A set of configurations of CS amplifiers using active loads is briefly presented in Fig. 2.3. Finally, the designation of CS comes from the fact that the source of the amplifier is not connected to the input and neither to the output, i.e., from the small-signal point of view, the source of the amplifier is a common node to both the input and the output. By using an active load instead of a resistor, a high-impedance output load can be realized without using excessively large devices or a large power supply voltage. Hence, for a given power supply voltage, a larger voltage gain can be achieved using an active load than what would be possible to achieve if a resistor is employed as the load instead [7]. An active load takes advantage of the nonlinear, large-signal transistor relationship between current and voltage to provide large small-signal resistances without large DC voltage drops. In detail, since the drain-source voltage equals the gate-source voltage, the MOS device is in saturation, and since the small-signal output resistance is often much higher than gm [1], the small-signal equivalent circuit of an active load follows Fig. 2.4.

In general, the active load is in fact part of a current mirror that supplies a bias current to drive the main transistor, as shown in Fig. 2.5, where an n-channel common-source amplifier is biased by a p-channel current mirror. A small-signal

(AC ground common to both input and output)

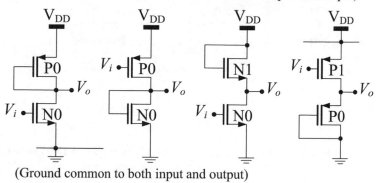

(Ground common to both input and output)

Fig. 2.3 Four possible configurations of common-source amplifiers with gate-drain loads

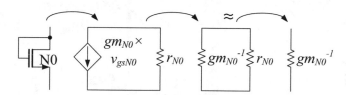

Fig. 2.4 Active load small-signal approximation

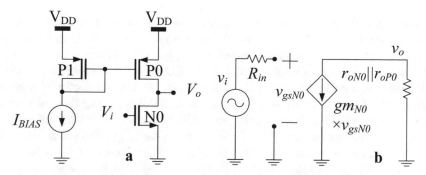

Fig. 2.5 Common-source amplifier with a current-mirror active load: (**a**) topology; (**b**) small-signal equivalent

equivalent circuit for a low-frequency analysis of the common-source amplifier is shown in Fig. 2.5, where V_i and R_{in} are the Thévenin equivalent components of the input voltage source. For further considerations, it is assumed that both transistors are operating in saturation, i.e., in the active region. The output resistance considers the parallel combination of the drain-to-source resistances of N0 and P0. Note that the voltage-controlled current source modeling the body effect is not included since the source is at a small-signal ground. Using small-signal analysis, the voltage gain of the configuration can be determined as shown in (2.5).

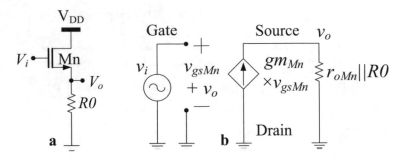

Fig. 2.6 Common-drain configuration: (**a**) circuit topology; (**b**) small-signal equivalent circuit

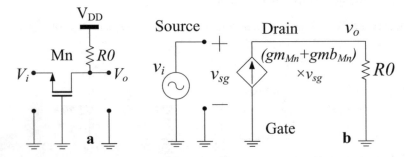

Fig. 2.7 Common-gate configuration: (**a**) circuit topology; (**b**) small-signal equivalent circuit

Another basic topology that is often referred in literature, and widely used in architectures, is the common-drain (CD) configuration, which definition follows the same principle as the previously detailed topology, yet, in this case, the gate is the common terminal to both input and output. The CD or source-follower stage is often implemented to operate as buffer since this circuit can provide a gain of approximately one. The gain expression can be given by (2.6) analyzing the small-signal equivalent circuit in Fig. 2.6, where the body effect is neglected for simplicity. When using single-well technology, the difficulty to avoid the body effect arises, since $vbs \neq 0$ V. Under these conditions, if an NMOS is considered, the bulk is connected to the ground, and (2.6) becomes (2.7). If $R0$ is sufficiently high, it is possible to define the gain of this topology as in (2.8). Finally, when the gate is common to the input and output, the common-gate (CG) configuration is definable, following the circuit in Fig. 2.7. In this case, the body effect is taken into account, r_o is neglected, and the gain is given by (2.9).

$$A_v = \frac{v_o}{v_i} = -gm_{N0}(r_{oN0} \| r_{op0}) \tag{2.5}$$

$$A_v = \frac{v_o}{v_i} = \frac{gm_{Mn} \times (r_{oMn} \| R0)}{1 + gm_{Mn} \times (r_{oMn} \| R0)} \tag{2.6}$$

$$A_v = \frac{v_o}{v_i} = \frac{gm_{Mn} \times \left(r_{o_{Mn}} \| R0 \| gm b_{Mn}^{-1}\right)}{1 + gm_{Mn} \times \left(r_{o_{Mn}} \| R0 \| gm b_{Mn}^{-1}\right)} \tag{2.7}$$

$$A_v = \frac{v_o}{v_i} \approx \frac{gm_{Mn}}{gm b_{Mn} + gm_{Mn}} \tag{2.8}$$

$$A_v = (gm b_{Mn} + gm_{Mn}) \times R0 \tag{2.9}$$

Introductory studies of active circuits in literature often devote a significant amount of time to standard single-ended amplifying configurations, e.g., the common-terminal configurations. Yet, the practical value of single-ended amplifier configurations is actually limited. The fact is that differential amplifiers dominate modern analog ICs. The first reason for this is that differential amplifiers apply gain not to one input signal but to the difference between two input signals. This means that a differential amplifier, naturally, attenuates noise or interference that is present in both input signals. Moreover, differential amplification suppresses common-mode signals, i.e., a DC offset that is present in both input signals will be removed, along with signal even harmonics. The gain will be applied only to the signal of interest, assuming that the signal of interest is not present in both inputs. This is particularly advantageous in the context of IC design because the intensive need for DC blocking capacitors is, this way, eliminated. There are, however, some aspects that need careful attention. First, there is evidentially the need for a higher number of components, which, from the cost perspective in general, does not represent a serious problem. Secondly, however, the symmetry is, in the vast majority of cases, of utmost importance and needs to be guaranteed. Fortunately for designers, recent IC technologies and fabrication techniques are very good at achieving consistency among components inside a chip, which is commonly referred as matching, which is conveniently also well modeled at simulation level. The description of a basic differential structure is presented further. The common-source differential pair with resistive loads is presented in Fig. 2.8. Although the sources of the devices that compose the amplifier are not actually grounded, they are grounded from the small-signal perspective, i.e., the sources are virtually grounded, since the current source provides a constant current [9]. If a differential signal v_d is applied at the inputs of the amplifier, i.e., $v_d/2$ is applied to the gate of N0a and $-v_d/2$ to the gate of N0b, it is possible to probe a signal between the drains of N0a and N0b, which is commonly referred as the output differential voltage, and has an amplitude that is twice the value of a single-ended voltage whether probed at the drain of N0a or N0b differential-pair transistors. The ration between the output differential voltage and v_d is the differential gain and, in the case of Fig. 2.8, can be given by (2.10) where gm is the transconductance of the differential pair, which is determined by (2.11). The analysis of fully differential blocks can be simplified, if they are completely symmetric and stimulated with a differential input, through the usage of the concept of "half circuit" [9]. This concept is used throughout this work, since the usage of this

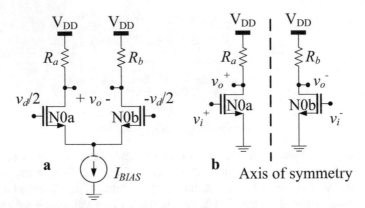

Fig. 2.8 Differential CS amplifier with resistive loads: (**a**) topology; (**b**) small-signal symmetry

concept greatly eases the analytical comprehension of the proposed topologies. Taking into account the differential pair with resistive loads shown in Fig. 2.8, and since the connection point of the sources of N0a and N0b suffers no fluctuations, this node is, as referred previously, an alternate current (AC) or virtual ground node; the complete circuit can be decomposed into two split branches, as illustrated in Fig. 2.8. Considering two common-source amplifiers, it is possible to describe the gain expressions of both branches as in (2.12), resulting in the differential gain shown in (2.13) [9]. In other words, it is possible to define a differential gain by analyzing only the gain of one half of the circuit. In fact, despite that the differential gain is twice the value of each branches of the single-ended gain, the input applied to one branch represents one half of the differential voltage.

$$A_v = \frac{v_o{}^+ - v_o{}^-}{v_i{}^+ - v_i{}^-} = \frac{v_o}{v_d} = -\text{gm} \times r_o \tag{2.10}$$

$$\text{gm} = \sqrt{2\mu_n C_{ox}\left(\frac{W}{L}\right) I_{D_{a,b}}} = \sqrt{\mu_n C_{ox}\left(\frac{W}{L}\right) I_{BIAS}} \tag{2.11}$$

$$\frac{v_o{}^+}{v_i{}^+} = -\text{gm} \times r_o \qquad \frac{v_o{}^-}{v_i{}^-} = -\text{gm} \times r_o \tag{2.12}$$

$$\frac{v_o{}^+ - v_o{}^-}{v_i{}^+ - v_i{}^-} = \frac{v_o}{v_d} - \text{gm} \times r_o \tag{2.13}$$

2.3   Performance Metrics of Amplifiers

The amplifier performance metrics, considered in the presented work, are briefly described in this section, prior to further considerations, in order to better understand the context of the development presented in this work.

2.3.1   Gain and Gain-Bandwidth Product

The establishing speed is considered through the GBW or unitary frequency of the amplifier, which corresponds to the frequency value where the voltage gain of the circuit is equal to 1 V/V, i.e., 0 dB, considering the frequency response illustrated in Fig. 2.9. The low-frequency gain, i.e., A_v, is importantly considered throughout this work. The stability of the amplifier, described further in this section, is evaluated through the phase margin (PM), in the context of non-inverting amplifier topologies, i.e., the PM is calculated through the sum of 180° to the phase value in frequency domain at GBW [9].

2.3.2   Figure of Merit

The energy efficiency of the amplifier is evaluated according to the figure of merit (FOM) defined in (2.14), i.e., the energy efficiency is evaluated through the ratio of the GBW and the current consumption, i.e., I_{DD}, for a given output load, i.e., C_L. The energy-efficient FOM, defined in (2.14), is widely established and commonly considered in related literature, with the purpose of comparison and ranking within the state of the art [10].

$$FOM = \frac{GBW \times C_L}{I_{DD}} \left[\frac{MHz \times pF}{mA} \right] \qquad (2.14)$$

Fig. 2.9 Asymptotic frequency-response of low-pass amplifiers

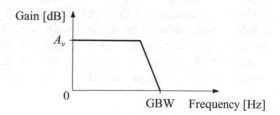

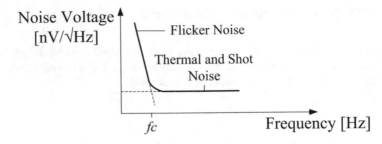

Fig. 2.10 Flicker noise and thermal and shot noise

2.3.3 Noise

Regarding the noise contribution of the amplifiers, the equivalent input-referred noise response, with emphasis on the effective noise contribution at the value of GBW, is considered. Moreover, both flicker noise and thermal noise are considered, which are illustrated and defined in Fig. 2.10. The flicker, or the $1/f$ noise, is often characterized by the corner frequency, i.e., f_c, situated between the region dominated by the low-frequency noise, i.e., the flicker noise, and the high-frequency noise, i.e., flat-band noise [9]. Regarding CMOS technologies, the thermal noise is predominant, when compared to the shot noise, which is often more preponderant in bipolar technologies since it regularly derives from a DC current flowing through a p-n junction.

2.3.4 Common-Mode Rejection Ratio

Another important aspect of a differential amplifier is its ability to reject a common-mode signal applied to both inputs. Commonly, in analog systems, signals are transmitted differentially, thus the capability of an amplifier to reject coupled noise in each line is very desirable. The common-mode rejection ratio (CMRR) evaluates how well an amplifier can reject signals which are common to both inputs and is defined as in (2.15), where A_{dm} is the differential gain, A_{cm} is the common-mode gain, and v_{cm} is the common-mode voltage, i.e., a common signal applied to both inputs [11]. Moreover, if an OpAmp with more than one stage is considered, the CMRR is determined by the input stage, leading to (2.16). In practical terms, the CMRR can be evaluated by subtracting the open-circuit frequency response, in terms of gain, by the corresponding response for small signals that come from the common-mode voltage, if the metrics are considered in dB. On another hand, the same corresponds to a quotient, if the metrics are evaluated in linear units.

$$CMRR = 20\log\left|\frac{A_{dm}}{A_{cm}}\right| = 20\log\left|\frac{v_o/v_i}{v_o/v_{cm}}\right| \qquad (2.15)$$

$$CMRR_{OPAMP} = CMRR_{input\ stage} \qquad (2.16)$$

2.3.5   Power Supply Rejection Ratio

Ideally, the power supplies, e.g., positive or negative and ground, are constant, so that the output voltage of an OpAmp depends only in the differential and common-mode input voltages provided to the OpAmp. However, in practice, the power supply voltages are not constant, and their variations contribute to the output of the amplifier, i.e., the ripple on the power supplies contributes with noise at the output, instead of neutral, as desirable.

The power supply rejection ratio (PSRR) evaluates how well an amplifier rejects noise or variations on the supplies or ground power buses. This parameter is of utmost importance in precision analog design and is defined in (2.17), as the ratio between the differential gain and the gain from the power supply ripple at the output with the differential input set to zero. The power supply ripple is v_{dd}.

$$PSRR = 20\log\left|\frac{v_o/v_i(v_{dd} = 0)}{v_o/v_{dd}(v_i = 0)}\right| \qquad (2.17)$$

The PSRR is commonly evaluated as the relation between the differential gain of an amplifier and the gain referred to the power supply, as opposed to the specific case of voltage regulators, for which the latter is sufficient [7], as shown in (2.18), and can be divided into two, i.e., the positive PSRR and the negative PSRR, considering the positive supply and negative supply contributions, respectively. In practice, the positive PSRR is defined as in (2.19) and the negative PSRR is defined as in (2.20). Still regarding voltage regulators, the ability to maintain a quiet voltage at the output, in the presence of noise from the voltage supply, is evaluated by superimposing a small signal onto the voltage supply and measuring the resulting variations at the output. The ratio of these two small signals results in the PSRR of the voltage regulator. In practical terms, the PSRR can be evaluated by subtracting the open-circuit frequency response, in terms of gain, by the corresponding response for small signals that come from the voltage supply source, if the metrics are in dB. As in the case of the CMRR, the same corresponds to a quotient, if the metrics are evaluated in linear units.

$$PSRR_{Voltage\ Regulator} = 20\log\left|v_{dd}/v_{reg}\right| \qquad (2.18)$$

$$PSRR^+ = \frac{v_o/v_i}{v_o/v_{dd}} \tag{2.19}$$

$$PSRR^- = \frac{v_o/v_i}{v_o/v_{ss}} \tag{2.20}$$

2.3.6 Slew Rate

When the input to an OpAmp circuit changes too quickly, the OpAmp is unable to maintain a virtual ground between the corresponding inputs. Hence, the OpAmp temporarily sees a large differential input. The slew rate (SR) of an OpAmp is the maximum slope of the output voltage as defined in (2.21), and, in general, this metric is evaluated in a closed-circuit buffer configuration, i.e., unity-gain configuration.

The positive SR can be different from the negative SR, as the rise time of the step response can also be different from the fall time of the step response [12]. The SR, which is a large-signal phenomenon, is generally determined by the maximum output current available to charge or discharge a capacitance and, in the case of multistage architectures, is not limited by the output but, rather, by the current sourcing/sinking capability [8].

$$SR = \left.\frac{\Delta v_o}{\Delta t}\right|_{max} = \left.\frac{dv_o}{dt}\right|_{max} \tag{2.21}$$

2.3.7 Settling Time

The settling time is the time required to settle the output within a given range, e.g., usually $\pm 0.1\%$ of the final value is considered, yet this limit can differ within diverse applications and fields, when a small signal is applied to the circuit. In other words, the settling time represents the time needed for the output of the OpAmp to reach a final value within a predetermined tolerance range, as represented in Fig. 2.11 [8].

2.3.8 Offset Voltage

In CMOS differential amplifiers, the offset voltage (VOS) is an important metric in the sense that the same may cause unexpected distortion and also cause implications in the circuitry ahead of the amplifier. The offset genesis is composed of two terms:

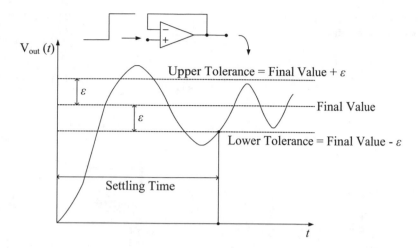

Fig. 2.11 Illustration of the settling time in the context of the transient step response of an OpAmp

the random offset and the systematic offset. The random offset is due to the geometrical mismatching and process-dependent inaccuracies. The systematic offset can be reduced to a value close to zero with a careful design, since this performance metric results from the design of the circuit and is present even when all the matched devices are physically identical. Unlike the random offset, which may be either positive or negative, a systematic offset will always have a known polarity.

In theory, when the input terminals of a given fully differential amplifier are connected together, the DC voltage values of both output branches are at two desired quiescent points, which ideally are equal. In a real differential amplifier, however, the output offset voltage is the actual difference between both quiescent points, which may not be zero. Moreover, when this offset voltage is divided by the differential voltage gain of the differential amplifier, the new metric is called the input offset voltage. Throughout this work, the offset voltage is defined as the difference between the actual DC output voltage values and the ideal DC output voltage values, expected when both input terminals are connected together, i.e., half of V_{DD}, as illustrated in Fig. 2.12.

2.3.9   Output Voltage Swing

The output swing voltage (OS) is defined as the maximum swing of the output node without generating a defined amount of harmonic distortion. In practice, the OS is determined by the difference between the positive supply and the negative supply voltages or ground minus the overdrive voltages of the transistors that drive the output node, i.e., transistors that form the output branches of the amplifier, as illustrated in Fig. 2.13.

Fig. 2.12 Illustration of the
systematic offset voltage

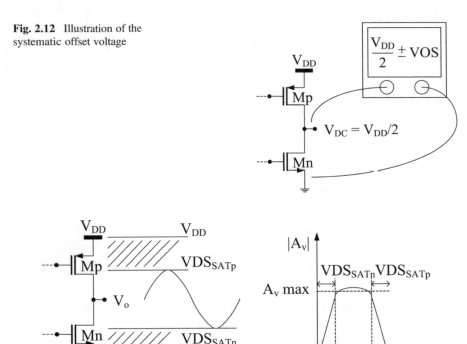

Fig. 2.13 Output voltage swing: (**a**) effect illustration; (**b**) OS versus the voltage gain

2.3.10   Stability and Frequency Compensation

Generically, in single-stage architectures, the stability of an amplifier is less problematic than in the case of their multistage counterparts. Stability and frequency
compensation are two related aspects which, in fact, have a dependency between
them: the frequency compensation is a mechanism to guarantee the stability of an
amplifier. The stability of an amplifier is defined, generically, as the ability to
produce bounded outputs in the presence of bounded inputs; an oscillating circuit
or state is defined as marginally stable, which is undesirable in the case of OpAmps.
In the particular case of single-stage OTAs, which are defined in detail in the next
section, the location of the dominant pole, which directly influences the stability of
the circuit, depends mainly on the value of a high impedance load at the output, e.g.,
capacitive load, and, with proper design considerations, no additional compensation
is needed, as addressed in the following paragraphs. The phase lag of the signal,
when going through the several stages of an OpAmp, determines the limit to the
useful gain at high frequencies. At the frequency where the phase lag exceeds 180°,
the open circuit gain must be dropped below unity. Otherwise, the feedback system
will become self-oscillating due to the fact that a negative feedback becomes, in fact,

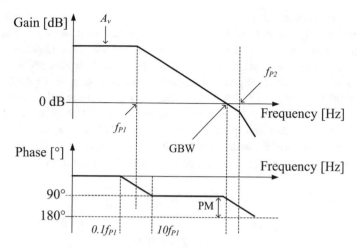

Fig. 2.14 Asymptotic frequency response of an OpAmp with gain and phase

a positive feedback. Moreover, a gain amplitude and phase margin must be adopted to obtain a response without peaking in the frequency domain or without overshoot in the time domain [6]. A generically desired frequency characteristic of an OpAmp in open-circuit configuration is presented in Fig. 2.14, which contains the gain and phase characteristic curves in a Bode diagram. Moreover, the PM concept is also illustrated in Fig. 2.14 and can be calculated as the difference between the phase value at GBW and 180°, representing, in practical terms, how far is an amplifier from diving into an oscillating state. In non-inverting topologies, the PM can be calculated by summing 180° to the output phase at GBW, as described in Sect. 2.3.1. The frequency response shown in Fig. 2.14 represents an amplifier with one dominant pole located at $\omega_{p1} = 2\pi f_{p1}$ and a limiting pole located at $\omega_{p2} = 2\pi f_{p2}$. Parasitic poles situated above the frequency of the limiting pole are disregarded. For a Butterworth pole position of the unity-gain feedback amplifier, with a flat frequency response and with an overshoot in the step response of 5% [6], the two pole frequencies of the open circuit must be separated at a distance of, at least, two times the low-frequency gain, as given by (2.22).

A larger separation between the two important poles is also allowed, increasing the PM to a value larger than 60°. For the majority of applications, however, a PM of 60° is sufficient, and enlarging this value represents a waste of bandwidth or current, and this is simply not needed. The procedure of relocating the poles in order to improve the PM, i.e., the process of obtaining a desired frequency response, is called frequency compensation, which in the case of multistage architectures mainly rely on additional circuitry, e.g., adding a Miller capacitor, in the best case, with direct impact on the area occupied by the amplifier. From (2.22), it follows that the GBW, which is the frequency where the gain crosses 0 dB as defined previously in Sect. 2.3.1, must be a factor of two below the limiting pole frequency, f_{p2}, as defined in (2.23). In the same way, the dominating pole frequency, f_{p1}, is situated a factor equal to the low-frequency gain, A_v, lower than the GBW, following (2.24). Moreover, if the amplitude characteristic between GBW and f_{p1} is not straight, but curved by a pole-zero doublet, the step response of the unity-gain feedback amplifier can be expected to have a slow settling overshoot or undershoot component. This is undesirable in high-speed amplifiers, i.e., amplifiers which need a fast and accurate settling after a step input signal. Therefore, a general-purpose OpAmp requires roughly 6 dB *per* octave, i.e., 20 dB *per* decade, slope between the dominant-pole frequency, and the GBW. As described in the next section, the proper design targeting a specific load is important in single-stage OTAs, since the load capacitance influences the GBW, hence the stability of the circuit [13], i.e., the ability to deliver a bounded output signal in the presence of a bounded input signal.

$$\frac{f_{p2}}{f_{p1}} \geq 2 \times A_v \tag{2.22}$$

$$f_{p2} \geq 2 \times \text{GBW} \tag{2.23}$$

$$f_{p1} = \frac{\text{GBW}}{A_v} \tag{2.24}$$

2.4 Operational Transconductance Amplifiers

In the last four decades, OTAs have been receiving considerable attention due to their usefulness and versatility in many filtering and signal processing applications, in which purely capacitive loads are driven. In fact, an OTA has often excellent high-frequency performance. Moreover, in some topologies, the gain can be made programmable through the DC current with large, or very large, adjustment ranges. Most OTA topologies can realize arbitrarily complex functions, and circuits are often simpler than what can be obtained with OpAmp counterparts. The circuit properties are, however, particularly process and temperature dependent in the majority of cases.

The basic electrical scheme and symbol of an OTA is shown in Fig. 2.15. An ideal OTA can be defined by (2.25). At this point, it is appropriate to define an OTA, generically, as a voltage-controlled current source with active gain, whereas

Fig. 2.15 CMOS operational transconductance amplifier

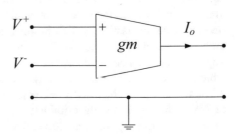

OpAmps are generically defined as voltage-controlled voltage sources. Therefore, an OTA is basically a voltage-to-current transducer. In other words, an OTA is essentially an OpAmp without the output stage. A generic OTA can also be defined as an amplifier where all nodes are of low impedance except the input and output nodes. The feature of an ideal OTA is that this type of amplifier has infinite input and output resistances.

In fact, a generic OTA can only drive purely capacitive loads; thus, revisiting Sect. 2.3.6, in this case the slew rate represents the fastest rate that the output current can charge and discharge a load capacitance. The SR expression can be derived from (2.26) and (2.27) and finally given by (2.28).

$$I_o = \text{gm} \times (V^+ - V^-) = \text{gm} \times v_i \qquad (2.25)$$

$$q = v_o \times C_L = \int (\text{gm} \times v_i) dt = \int (i) dt \qquad (2.26)$$

$$\frac{dv_o}{dt} = \frac{i}{C_L} = \frac{\text{gm}}{C_L} \times v_i \qquad (2.27)$$

$$\text{SR} = \left.\frac{dv_o}{dt}\right|_{\max} = \left.\frac{i}{C_L}\right|_{\max} \qquad (2.28)$$

The approaches proposed in this work are implemented as modifications of the basic symmetrical CMOS operational transconductance amplifier, shown in Fig. 2.16 [13], which provides the basis for the developed topologies. The symmetrical CMOS OTA is the most suitable circuit for proof of concept, since a minimum number of stacked transistors are used. The following paragraphs provide a description of the symmetrical CMOS OTA.

The circuit shown in Fig. 2.16 comprises a differential pair, i.e., N0 pair, in common-source configuration, biased in current, and loaded with active current mirrors, i.e., P1 and P0 pairs. It is important to consider that the gain of the symmetrical CMOS OTA is only dependent on the small-signal transconductance of the differential-pair devices, on the current-mirroring factor B and on the output resistance, r_o, as shown in (2.29). Furthermore, the GBW is directly proportional to B and the first two terms and minimized by the capacitive load, as shown in (2.30). The advantages of the symmetrical CMOS OTA rely on better systematic offset; better OS, which can be given by (2.31); and better common-mode rejection ratio specifications, when compared to the classic two-stage Miller-compensated OTA [9]. Nevertheless, the gain defined in (2.29) is dependent only in the transconductance of the differential pair, on the output resistance, and on the current-mirroring factor, if a static current source is employed to bias the differential pair. This leaves room to the principle of the following work, i.e., the replacement of the static current source by an alternative structure, with the purpose of enhancing the low-frequency gain and the energy efficiency of the OTA.

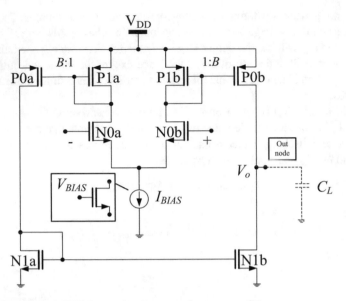

Fig. 2.16 Symmetrical CMOS operational transconductance amplifier

$$A_{\mathrm{v}} = \mathrm{gm}_{\mathrm{N0}} \times B \times r_{\mathrm{o}} \qquad (2.29)$$

$$\mathrm{GBW} = \frac{\mathrm{gm}_{N0} \times B}{2\pi \times C_{\mathrm{L}}} \qquad (2.30)$$

$$\mathrm{OS} \approx V_{\mathrm{DD}} - 2 \times \mathrm{VDS}_{\mathrm{SAT}} \qquad (2.31)$$

2.5   The Cascode Amplifier Topologies

This section presents and discusses a technique to increase the gain of a single-stage amplifier, named cascode. The motivation for using cascode configurations to increase the gain of an amplifier relates also to the vast number of non-buffered applications in which two-stage operational amplifiers, e.g., two-stage CMOS with Miller compensation, have poor performance. Performance limitations of the two-stage CMOS with Miller compensation include low gain, limited stable bandwidth, and a poor PSRR [14]. Therefore, the cascode technique is primarily applied when high gains are needed. However, with the tendency of low power supplies, this technique can suffer severe drop backs in the near future, due to transistor limited stacking.

2.5.1   Basic Cascode Stages

The basic cascode configuration comprises only two transistors, as illustrated in Fig. 2.17. This two-device structure consists of a common-source-connected transistor feeding into a common-gate-connected transistor. Historically, the cascode designation has been brought through times since the days when scientists used vacuum tubes and results from the grammatical contraction of two words: cascaded and cathode, hence cascode. The gain expression, given by (2.32), can be derived from the transconductance of N0 times the output resistance, which is affected by N1, as shown in (2.33). This configuration is commonly called telescopic-cascode stage [7]. The frequency response of this topology is dominated by a single pole, which is calculated as in (2.34), if the gate-drain capacitance *cgd* of N1 is comparable with an output load capacitance, C_L, that usually dominates the pole [7, 15].

Following this approach, it is reasonable to increase the complexity of the telescopic-cascode stage to an implementation that is more important in practical terms. In fact, the telescopic-cascode amplifier described further is basically a symmetric implementation of the double telescopic-cascode stage, shown in Fig. 2.18, with properly adopted biasing, i.e., including now a p-channel MOSFET (PMOS) nonideal cascode current source. The gain of this stage, following the same analysis of the previous circuit, is given by (2.35), while the output resistance is given by (2.36), with minor simplifications in gm, i.e., both NMOS devices are sized to have equal transconductance and small-signal output resistance. The same is true in the case of the NMOS devices. The pole frequency is given in (2.37), while, to obtain the OS, it is enough to consider the difference between the supply voltage and the sum of the overdrive voltages of P0, P1, N1, and N0; therefore, the OS is fairly limited. Also, in order to eliminate the Miller effect, it is desirable to have R0 rather large, and C0 sized to present low impedance, in terms of AC signals [11].

$$A_\mathrm{v} = -\mathrm{gm}_{\mathrm{N0}} \times r_\mathrm{o} \approx -\mathrm{gm}_{\mathrm{N0}}\mathrm{gm}_{\mathrm{N1}}r_{\mathrm{oN1}}r_{\mathrm{oN0}} \qquad (2.32)$$

Fig. 2.17 Telescopic-cascode stage with ideal current source

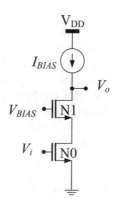

Fig. 2.18 Double
telescopic-cascode stage
with cascode PMOS current
source

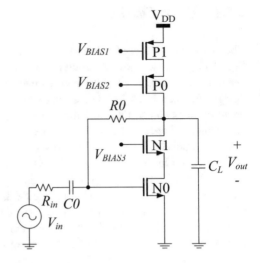

$$r_{\rm o} = \left(gm_{\rm N1} r_{\rm oN1} + 1\right) \times r_{\rm oN0} + r_{\rm oN1} \approx gm_{\rm N1} r_{\rm oN1} r_{\rm oN0} \tag{2.33}$$

$$f_{-3\,\rm dB} \approx \frac{1}{2\pi \times \left(C_{\rm L} + cgd_{\rm N1}\right) \times r_{\rm o}} \tag{2.34}$$

$$A_{\rm v} \approx -\frac{\left(gm_{\rm N1}(r_{\rm oN})^2\right) \| \left(gm_{\rm P0}(r_{\rm oP})^2\right)}{\frac{1}{gm_{\rm N0}}} = -gm_{\rm N0} r_{\rm o} \tag{2.35}$$

$$r_{\rm o} \approx \left(gm_{\rm N1}(r_{\rm oN})^2\right) \| \left(gm_{\rm P0}(r_{\rm oP})^2\right) \tag{2.36}$$

$$f_{-3\,\rm dB} \approx 1/(2\pi \times \left(C_{\rm L} + cgd_{\rm N1} + cdg_{\rm P0}\right) \times r_{\rm o}) \tag{2.37}$$

In order to reduce the impact of stacking a large number of transistors across a lower voltage power supply, it is possible to replace the common-gate NMOS transistor of the telescopic-cascode stage by a PMOS device in a common-gate configuration in such way that the basic behavior principles and analytics remain unchanged. The configuration shown in Fig. 2.19 comprises an n-channel input transistor, but a p-channel transistor is used for cascode, i.e., common-gate, transistor. This configuration is usually named as the folded-cascode. This topology is denominated as folded, in the sense that the PMOS reverses the direction of the signal flow back toward the ground, i.e., the signal current gm $\times v_{\rm i}$ is folded down and made to flow into the source terminal of P0. The gain of this topology follows the expression of its telescopic counterpart, i.e., the gain is similarly given by (2.32).

When incorporating a differential OpAmp, the folded cascode can provide greater OS than the telescopic cascode and increases the common-node input range as well,

Fig. 2.19 Folded-cascode
stage

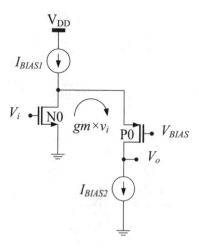

becoming more independent in terms of applied DC voltage [14]. However, the folded usually consumes more power, since the drain bias currents for N0 and P0 are drawn in parallel from the supply, whereas in the telescopic cascode, the same DC drain current is shared by both transistors. The power dissipation of this structure is approximately equal to $V_{DD} \times 2I_{BIAS1}$. In the next subsections, a set of amplifiers using the previous amplification stages are presented.

2.5.2 Telescopic-Cascode Amplifier

The telescopic-cascode amplifier (TCA) makes use of the basic principle of the telescopic-cascode stage presented in the previous section. This amplifier allows a gain comparable to a two-stage OpAmp [7]. The topology is shown in Fig. 2.20. The main advantages of this amplifier, when compared to a two-stage Miller-compensated amplifier, are as follows: first, the bias is carried out with a single bias current source, and second, this circuit is characterized by a high speed with less Miller effect [7]. The disadvantages are also two: first, the positive swing is limited by two devices while the negative swing is limited by two overdrives plus a threshold voltage, and therefore the OS is relatively low [14]; on another hand, the common-mode input range is also limited by the threshold voltage of N1, as illustrated in Fig. 2.21, if in closed-circuit configuration, i.e., with the output node connected to the gate of N0b. The common-mode input range can be maximized by minimizing the overdrive of N1, yet the first is always less than the V_T of N0. These aspects are better if a fully differential implementation is considered. In order to improve the OS, wide-swing PMOS current mirrors can be employed.

The open-circuit gain of the TCA can be given by the contributions of the pair transconductance and the output resistance. The gain expression is shown in (2.38), and the output resistance is given by (2.39) [9]. It is at this point important to

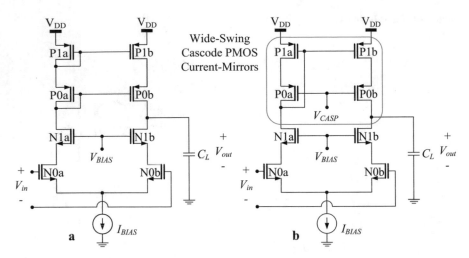

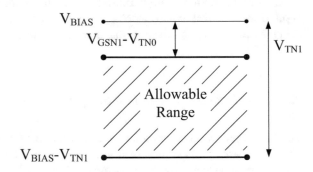

Fig. 2.20 Telescopic-cascode amplifier: (**a**) classic cascode PMOS; (**b**) wide-swing cascode PMOS

Fig. 2.21 Telescopic-cascode common-mode input range limitations in unity-gain configuration

consider also that the transconductance of the amplifier is equal to the transconductance of the differential pair since cascoding, by itself, has no impact on the transconductance of the amplifier. Overall, the TCA provides, in general, a good tradeoff between gain, power consumption, speed, and noise; but the output swing of this architecture is limited, and this topology is fairly difficult to use in closed circuit, e.g., to implement a unity-gain buffer, since shorting the input and output is difficult due to the intrinsically low common-mode input range.

$$A_v = gm_{N0} \times \left(\frac{gm_{P0} \times gm_{N1} \times (r_{oP1} \times r_{oP0} \times r_{oN0} \times r_{oN1})}{gm_{P0} \times r_{oP1} \times r_{oP0} + gm_{N1} \times r_{oN0} \times r_{oN1}} \right) \qquad (2.38)$$

$$r_o = ((gm_{P0} \times r_{oP0}) \times r_{oP1}) \| ((gm_{N1} \times r_{oN1}) \times r_{oN0}) \qquad (2.39)$$

2.5.3 Mirrored-Cascode Amplifier

In recent literature, the topology presented in this subsection is often not explored. However it is worth mentioning, since this circuit has many important aspects to be considered. The mirrored-cascode amplifier (MCA), shown in Fig. 2.22, is, in a way, a combination of the folded-cascode amplifier with the symmetrical CMOS OTA. The MCA makes use of current-mirrors to drive the output node, improving the gain by the mirroring factor B, as in the case of the symmetrical CMOS OTA. To improve the OS, wide-swing PMOS current mirrors can be employed in the cascode topologies, as shown in Fig. 2.22. Moreover, higher-output impedance can be generated with this topology, due to the current mirrors. The gain is this way proportional to the B mirror factor as shown in (2.40).

$$A_v \propto B \times gm_{N0} \times r_o \qquad (2.40)$$

An optimum input common-mode range is achieved with this topology [7]. Moreover, only four transistors in series improve the output swing of the amplifier. The major drawback, relative to the telescopic-cascode amplifier and the folded-cascode amplifier, which is described in the next subsection, is the higher power consumption that this topology presents, precisely due to the current-mirroring technique. The power dissipation is approximately equal to $V_{DD} \times 2(1 + B) \times I_{BIAS}$, which is, expectedly, greater than the previous topology by a factor of B.

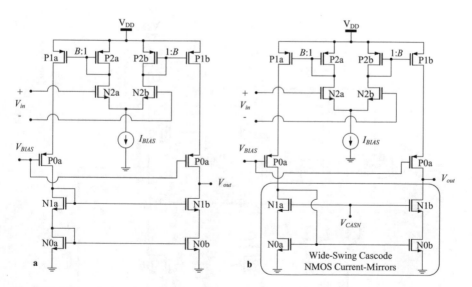

Fig. 2.22 Mirrored-cascode amplifier: (a) classic cascode NMOS; (b) wide-swing cascode NMOS

2.5.4 Folded-Cascode Amplifier

Recalling that a cascode stage can be folded, this principle gives birth to the classic folded-cascode amplifier (FCA), which topology is shown in detail in Fig. 2.23. By using a folded cascode, the problem of headroom is alleviated. On another hand, by using PMOS transistors, the nondominant pole is at a lower frequency due to the fact that the PMOS has lower transit frequency [9], which can have implications in terms of GBW, as described further.

With a deeper insight on the FCA, it is possible to acknowledge that the output resistance of this single-stage amplifier is increased by the intrinsic gain of the cascode devices N1a,b and P0a,b, comparatively to the basic two-stage OpAmp with a common-source second stage. Moreover, the transconductance is simply the gm of the input pair; thus, the voltage gain is comparable to that of the two-stage OpAmp. The transconductance of the FCA is given by gm_{N2}, while the output resistance is composed of two contributions: the portion related to the PMOS devices above the output node and the portion related to the NMOS devices below the output node, as illustrated in Fig. 2.23. The expression of the output resistance is composed of two contributions: the first contribution is expressed in (2.41), and the second contribution is presented in (2.42). Hence the output resistance is given by (2.43). Finally, the gain of this topology is given by (2.44).

$$r_{\mathrm{oUP}} = gm_{\mathrm{P0}} \times r_{\mathrm{oP0}} \times (r_{\mathrm{oP1}} \| r_{\mathrm{oN2}}) \tag{2.41}$$

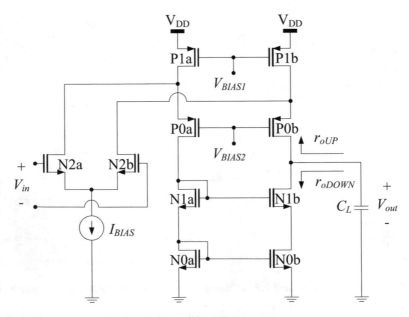

Fig. 2.23 Conventional folded-cascode amplifier (FCA)

$$r_{o_{DOWN}} = gm_{N1} \times r_{o_{N1}} \times r_{o_{N0}} \tag{2.42}$$

$$r_o = r_{o_{UP}} \| r_{o_{DOWN}} \tag{2.43}$$

$$A_v = gm_{N2} \times r_o \tag{2.44}$$

The dominant pole of the FCA is associated with the output node [6]. Considering that the load capacitance provides the frequency compensation, an increase of the latter directly lowers the GBW of the amplifier without, however, compromising the PM. On the other hand, the nondominant pole is associated with the cascode devices P0a and P0b. This pole is, in general, located near the transit frequency of the PMOS but also close to the unity-gain frequency of the amplifier; thus, this pole must be carefully designed to avoid limitations in terms of GBW. The amplitude Bode diagram, with the pole locations of the FCA, is shown in Fig. 2.24 [6].

A high-speed and a high-voltage gain are kept, in comparison with the TCA, with modest deterioration, yet a higher OS can be achieved, with a better possibility of successfully connecting the output with the input in closed-circuit configurations. Moreover, a high PSRR value can be achieved, since there is no pole splitting. In opposition, the most important drawback of this topology is related with the need for higher bias current with direct implications in power consumption. Moreover, an increase in the circuit size is expected. A comparison between the telescopic-cascode, mirrored-cascode, and classic folded-cascode amplifiers is presented in Table 2.1, in qualitative terms, summarizing the most important topologies presented at this point. A summary of the maximum output swing expressions of these same topologies is presented in Table 2.2, emphasizing the future limitations, related with the tendency to have lower supply voltages, with consequences in practical implementation terms. At this point, it is important to revisit (2.31), where the OS of the symmetrical CMOS OTA is defined. In fact, it is possible to acknowledge that the latter has a clearly better OS, due to lack of stacked devices in the output branches. When compared to the three cascode topologies described in the previous three

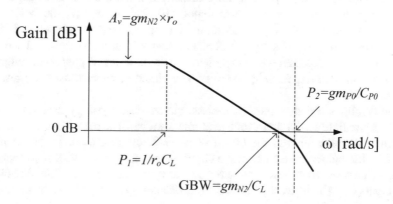

Fig. 2.24 Amplitude asymptotic Bode diagram of the FCA with GBW and pole expressions

Table 2.1 Comparison between telescopic-cascode, mirrored-cascode, and folded-cascode amplifier topologies

Topology	Gain	Output swing	Speed	Power dissipation	Noise	Lasting appeal with lower voltage supplies
Telescopic	High	Low	Moderate	Low	Low	Low
Mirrored	High	Moderate	Moderate	High	Moderate	Moderate
Folded	High	Moderate	Moderate	Moderate	Moderate	Moderate

Table 2.2 Output swing expressions of telescopic-cascode, mirrored-cascode and folded-cascode amplifiers

Topology	Maximum positive output signal swing	Maximum negative output signal swing	Maximum available output signal swing
Telescopic	$V_{DD} - 2 \times VDS_{SAT}$	$V_{SS} - 3 \times VDS_{SAT}$	$V_{DD} - V_{SS} - 5 \times VDS_{SAT}$
Mirrored	$V_{DD} - 2 \times VDS_{SAT}$	$V_{SS} - 2 \times VDS_{SAT}$	$V_{DD} - V_{SS} - 4 \times VDS_{SAT}$
Folded	$V_{DD} - 2 \times VDS_{SAT}$	$V_{SS} - 2 \times VDS_{SAT}$	$V_{DD} - V_{SS} - 4 \times VDS_{SAT}$

subsections, the maximum OS limitation is reduced, at least, by a factor of two, i.e., the OS of the symmetrical CMOS OTA represents, in the worst case, twice the value of the cascode topologies. This has utmost importance if lower voltage supplies are considered.

2.5.5 Gain Enhancement Techniques for the Cascode Architectures

With the objective of overcoming the limitations in terms of gain of single-stage architecture, important techniques have been developed to increase the gain of cascode amplifiers and are presented in almost all literature, as well as indispensable upgrades to the classic cascode topologies [10, 14]. In reality, the basic gain-boosting techniques rely in the addition of supplementary amplifiers, which in general are composed of simple differential stages with active loads. Therefore, additional area is expected with immediate penalty in terms of power consumption. The basic gain-boosting technique applied to a telescopic-cascode stage is presented in Fig. 2.25 [16].

The amplifier added to the basic telescopic-cascode topology uses a negative-feedback circuit to set the drain voltage of transistor N0. The main amplifier gain can be almost doubled through the use of negative-feedback circuit. In the circuit of Fig. 2.25, the output impedance of the circuit is increased when V_x is made equal to V_{REF}, which results in general in a stable output voltage, i.e., less systematic offset at the output [17]. The increase in the output resistance, given by (2.45), is reflected in

Fig. 2.25 Cascode gain
stage with gain
enhancement

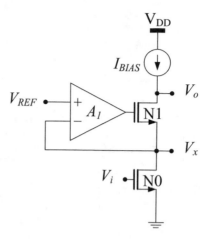

the overall gain of the stage which is, therefore, given by (2.46). Moreover, this gain-boosting structure also moves the nondominant pole to higher frequencies, since, in a first approach, the dominant pole, which frequency is given by $f_{p1} \approx 1/(2\pi \times r_o \times C_L)$, is dominated by the load capacitance, while the nondominant pole is given by $f_{p2} \approx 1/(2\pi \times r_x)$, with $r_x \approx 1/(gm_{N1} \times (1 + A_1))$, hence proportional to the A_1 gain.

$$r_o = (gm_{N1}r_{oN1}(A_1 + 1) + 1) \times r_{oN0} + r_{oN1} \approx gm_{N1}r_{oN1}A_1 r_{oN0} \qquad (2.45)$$

$$A_v = gm_{N0} \times r_o \approx gm_{N0}gm_{N1}r_{oN0}r_{oN1}A_1 \qquad (2.46)$$

One way to increase the gain of the FCA without cascading additional stages is to implement additional amplifiers in negative-feedback circuits, as shown in Fig. 2.26, to increase the resistance seen by the drain of each cascode transistor. In fact, to find the output resistance looking into the drain of P0b, it is enough to follow (2.48), whereas the output resistance looking from the drain of N1b can be determined by following (2.50). In order to determine the gain of the topology shown in Fig. 2.26, the original expression remains unchanged, i.e., the gain is given by $gm_{P2} \times r_o$, while, in this case, the output resistance is given by the parallel of (2.47) and (2.49). Thus, the output resistance of the topology illustrated in Fig. 2.26 is given by (2.51), and the gain can be given by (2.52).

$$r_{oUP} = r_{oP1} + r_{oP0}(1 + (gm_{P0}(A_{vA1b} + 1) + gmb_{P0})r_{oP1}) \qquad (2.47)$$

$$r_{oUP} \approx (A_{vA1b} + 1)(gm_{P0}r_{oP1})r_{oP0} \qquad (2.48)$$

$$r_{oDOWN} = (r_{oP2}\|r_{oN0}) + r_{oN1}(1 + (gm_{N1}(A_{vA2b} + 1) + gmb_{N1})(r_{oP2}\|r_{oN0})) \qquad (2.49)$$

$$r_{oDOWN} \approx (A_{vA2b} + 1)(gm_{N1}(r_{oP2}\|r_{oN0}))r_{oN1} \qquad (2.50)$$

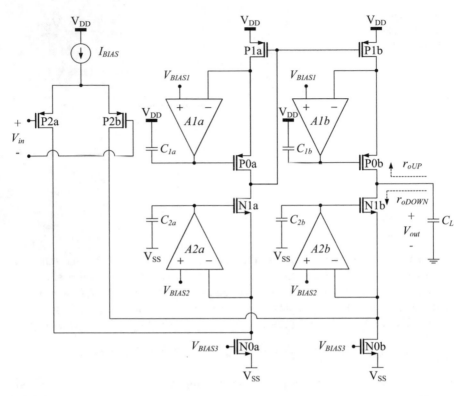

Fig. 2.26 Folded-cascode amplifier with active-cascode gain enhancement auxiliary amplifiers

$$r_o = r_{o_{UP}} \| r_{o_{DOWN}} \qquad (2.51)$$

$$A_v = gm_{P2} \times r_o \qquad (2.52)$$

Through the equations above, it is possible to acknowledge that the gain enhancement in the FCA, shown in Fig. 2.26, is not due to the auxiliary amplifiers that drive
the gates in the left branch of the circuit. However, these are indispensable to reduce
the systematic offset, if carefully designed equal to the ones in the right branch,
which directly influence the gain. Moreover, using desirably identical auxiliary
amplifiers to drive the gates of both P0a and N1a balances the two signal paths,
until the differential signal from the differential pair is converted into a single-ended
one by the current-mirror. The same is true for the auxiliary amplifiers that drive
the gates of both P0b and N1b [14]. Nevertheless, a problem of the presented topology
arises directly from the usage of auxiliary amplifiers, and it relies on potential
instability in the feedback circuits around them. In order to reduce the risk of
instability, capacitors C1a and C1b are placed between the outputs of the auxiliary
amplifiers and V_{DD} node, while C2a and C2b are placed toward V_{SS} node. In
principle, this can contribute to a significant increase of circuit area. The need for
these capacitors is related with the fact that the capacitances at the gates of P0a, P0b,

N1a, and N1b can be quite small, therefore providing small load capacitances to the auxiliary amplifiers, and, if not properly compensated, they can be dominated by parasitic effects that may vary significantly over processing and fabrication, ultimately leading to important stability issues [14].

2.6 The Recycling Folded-Cascode Amplifier

The recycling folded-cascode amplifier (RFCA) is the first nonconventional topology addressed in this work and described in detail in the following paragraphs. The RFCA makes use of devices inside the signal path, resulting in an enhanced transconductance, gain, and slew rate and achieving a FOM of 939 MHz × pF/ mA as shown in [1]. The simplified electrical scheme of the work described in [1] is presented in Fig. 2.27. The amplifier proposed in [1] delivers a clear enhancement of performance over the conventional FCA, as described in this section.

This performance enhancement is achieved by using previously idle devices in the signal path. Moreover, the input-referred noise and offset analyses presented in [1] show that the modifications have no adverse effects on these performance metrics. The modifications proposed in [1] are intended to use M3 and M4 as driving transistors, contributing with small-signal current to the folding node. The cross-coupled connections of these current mirrors ensure the small-signal currents added

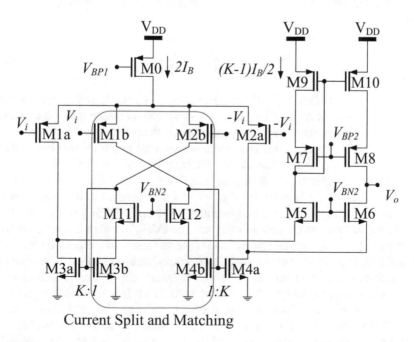

Fig. 2.27 The recycling folded-cascode amplifier (RFCA) proposed in [1]

at the sources of M5 and M6 are in phase. The transistors M11 and M12 ensure the matching by maintaining the drain potentials of M3 and M4 pairs. In order to present the enhancements proposed in [1] quantitatively, all devices are assumed to operate in the saturation region, and it is useful to consider the output resistance of the classic FCA, given by (2.53), and the corresponding voltage gain given by (2.54). The RFCA output resistance and gain expressions are, correspondingly, given by (2.55) and (2.56).

$$r_{o_{FC}} = (gm_{M6}rds_{M6}(rds_{M2}\|rds_{M4}))\|gm_{M8}rds_{M8}rds_{M10} \qquad (2.53)$$

$$A_{v_{FC}} = gm_{M1} \times r_o \qquad (2.54)$$

$$r_{o_{RFC}} = (gm_{M6}rds_{M6}(rds_{M2a}\|rds_{M4a}))\|gm_{M8}rds_{M8}rds_{M10} \qquad (2.55)$$

$$A_{v_{RFC}} = (gm_{M1a} \times (1 + K)) \times r_o \qquad (2.56)$$

The gain is mainly improved due to the current-mirroring factor K, which allows the gain to be easily enhanced by 10 dB, with respect to the classic FCA using the same technology node, as demonstrated in [1]. Regarding the phase margin of this topology, the proper choice of K is of major importance. Hence the application of the amplifier must be considered when dealing with this aspect. In fact, for high-speed applications, K can be chosen such that $\omega_{p3} > 3 \times$ GBW, where ω_{p3} represents the pole-zero pair of the OTA, setting a limitation to K as in (2.57). The authors of [1] define a value between 2 and 4, as a reasonable range for maximizing the phase margin.

$$K < \sqrt{\frac{gm_{3b} \times C_L}{3 \times gm_{1a} \times cgs_{3b}} - 1} \qquad (2.57)$$

The RFCA amplifier is firstly improved in [19] and, later on, revisited and improved in the work proposed in [20], in which a gain-boosting and a phase margin enhancement, by means of increasing the output resistance of the cascode pairs M5, M6 and M7, M8, are proposed. The topology presented in [20], which improves the FOM of the RFCA to a value of 987 MHz $\times$ pF/mA, is shown in Fig. 2.28. Furthermore, to improve the PM, transistors Mx and Mz that operate in triode region, cancel out the nondominant poles at current-mirror nodes, i.e., gates of M3a and M4a, bringing the first nondominant poles to the folding nodes, i.e., drains of M3a and M4a, as in the case of conventional FCA. The gain-boosting principle proposed in [20] relies on increasing the output resistance of the cascode pairs M5, M6 and M7, and M8 as shown in Fig. 2.27. The enhanced resistance seen from the drain of M6, if the input branch is ignored for simplicity of analysis, is given by (2.58), where A_{M20} is the gain of transistor M20 which is connected in a common-source configuration and operates as an auxiliary amplifier. The same analysis can be applied to the other three cascode devices. The auxiliary amplifiers are implemented using low-threshold transistors M14, M15, M20, and M21, while the transistors

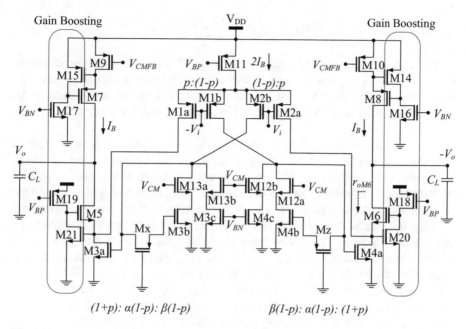

Fig. 2.28 The improved recycling folded-cascode amplifier with gain-boosting [20]

M16, M17, M18, and M19 emulate their current load. The total output resistance of the architecture is given by (2.59), where A_{M14} is the gain of the auxiliary common-source amplifier M14 and γ can be depicted by (2.60) [20]. The parameters α, β, and p are the mirroring factors represented in Fig. 2.28.

$$r_{oM6} = (\mathrm{gm_6 rds_6}(1 + A_{M20}) + 1)\mathrm{rds}_{4a} \tag{2.58}$$

$$r_o = \gamma(\mathrm{gm_6 rds_6}(1 + A_{M20}) + 1)\mathrm{rds}_{4a} \| (\mathrm{gm_8 rds_8}(1 + A_{M14}) + 1)\mathrm{rds}_{10} \tag{2.59}$$

$$\gamma = \left(\frac{(\alpha + \beta)(1 - p) + (1 + p)}{-(\alpha + \beta)(1 - p) + (1 + p)}\right)\mathrm{rds}_{2a} \| \left(\frac{(\alpha + \beta)(1 - p) + (1 + p)}{(1 + p)}\right)\mathrm{rds}_{4a} \tag{2.60}$$

It is assumed that the transconductance of the work in [1] remains unchanged by adding gain-boosting devices. Hence, the total gain of the proposed amplifier is expressed in (2.61), where r_o is given by (2.59), $\mathrm{Gm_{IRFCA}}$ is depicted according to (2.62) as proposed in [19], and Gm represents the transconductance of conventional FCA.

$$A_{\mathrm{VIRFCGB}} = \mathrm{Gm_{IRFC}} \times r_o \tag{2.61}$$

$$\mathrm{Gm_{IRFC}} = \left(p + \frac{(1+p)}{\alpha}\right) \times \mathrm{Gm} \tag{2.62}$$

A dissimilar gain-boosting technique is proposed in [21], which takes advantage of body effect to improve the gain of the RFCA. The current consumption of the work proposed in [21] is low, i.e., less than 1 mA. Yet, the main drawback of the work proposed relies on the GBW and, therefore, the speed, i.e., settling time, of the OTA.

The work presented in [22] describes a remarkable solution to improve the gain of a single-stage amplifier, without adding a significant penalty in power consumption. In the approach proposed in [22], a negative resistance structure is used in parallel with the output, improving the gain of the original topology, taking into account the maximization of the output swing. The circuit improvement proposed in [22] started initially from an n-channel MOSFET folded cascode and applying the proposed technique. This approach, however, reveals to be quite unsatisfactory in terms of GBW when compared with the solution in [1], achieving a value of around 12.5 MHz, compared with the GBW value of the work proposed in [1], i.e., a GBW of 134.2 MHz. The classic FCA topology has also been considered and improved in [23], through a gain-boosting technique which considers negative feedback. This approach, however, pays a considerable price in terms of power consumption. Still, regarding negative feedback, a solution that is suitable to be implemented in high-speed and high-resolution ADCs is presented in [24], yet still suffering from high-current consumption, i.e., 5.4 mA, yet with an interesting FOM of 1071.5 MHz × pF/mA, when considering the existing state of the art.

Single-stage OTAs with simultaneously high-gain and high-output swing have already been proposed either based on replica-amp gain enhancement techniques [25] or negative-feedback (gain-boosting) techniques [26, 27]. Unlike conventional techniques such as cascode, which increases the gain by increasing the output resistance, in [25] the replica-amp technique increases the gain by matching the main and the auxiliary replica amplifiers. However, the complexity of this replica-amp technique is very high, and the energy efficiency due to the need of an auxiliary structure is low. On the other hand, gain-boosting techniques thought negative feedbacks as proposed in [26, 27] are based on compound common-source/common-gate structures, i.e., cascode structures which, again, strongly limit the output swing of single-stage amplifiers. The work proposed in [28] corresponds to the implementation of a gain-boosting technique with the use of positive feedback, which is carried out by active devices, in a conventional FCA. The technique proposed in [28] does not increase the number of nodes in the initial structure of the OpAmp or the pole-zero doubling in the frequency response of the OpAmp. Therefore, applying this technique to the conventional FCA does not increase the settling time of the circuit in closed-circuit configuration. Moreover, the output voltage swing of the OpAmp proposed in [28] remains equal to that of the conventional topology, i.e., the output swing is limited by $V_{DD} - 4 \times \mathrm{VDS_{SAT}}$.

Table 2.3 Summary of state-of-the-art conventional and improved folded-cascode amplifiers

References	Gain [dB]	I_{DD} [mA]	PM [°]	GBW [MHz]	FOM [MHz × pF/mA]	Load [pF]	Tech [nm]	Year
[1]	60.9	0.800	71	134.2	939.4	5.6	180	2009
[19]	54.0	1.000	62	965.0	965.0	1.0	180	2010
[20]	85.6	1.000	67	987.0	987.0	1.0	180	2015
[21]	72.0	0.650	70	159.0	536.5	2.2	180	2007
[22]	84.0	0.085	81	12.5	146.0	1.0	180	2013
[23]	94.9	3.670	82	414.0	225.0	2.0	130	2011
[24]	91.5	5.000	62	714.5	1071.5	7.5	130	2012
[28]	67.0	2.170	67	920.0	211.0	0.5	180	2011

A summary of the works presented in [1, 19–24, 28], representing the most relevant state of the art in the field of folded-cascode-based amplifier topologies, is presented and compared in Table 2.3, showing several topologies which are less efficient from the energetic point of view, e.g., the works proposed in [21, 22, 28].

2.7 Dynamic CMOS Amplifiers

The first dynamic CMOS amplifiers are proposed by Copeland and Rabaey in 1979, in the work published in [29]. One year later, in 1980, Bedrich J. Hosticka proposed a new set of dynamic CMOS amplifiers and experimentally demonstrated the advantages of dynamic biasing [30]. The basic idea is that the bias current of such an amplifier is not constant as in the case of a static amplifier, i.e., static biasing, but changed during the amplifying operation, i.e., the dynamic biasing principle, and interrupted afterward to improve the power consumption of the circuit.

The property of having a dynamic current is of paramount importance in switched-capacitor (SC) circuits, in which the amplifier can be biased in strong inversion in the beginning of a given amplifying/integrating clock phase, i.e., starts operating with high slew rate and GBW, and then continuously reduces the current toward the weak-inversion region, i.e., subthreshold, until the power consumption is practically eliminated, thus maximizing both the finite gain and the output swing of the amplifier. The behavior of a static CMOS inverter, in terms of low-frequency gain, and GBW progression as functions of the bias current are illustrated in Fig. 2.29 [30], where it is possible to verify that, on one hand, the voltage gain is higher with a lower biasing current and, on the other hand, the GBW is proportional to the current consumption of the circuit.

In the work proposed in [30] and later on developed in [31], a new set of dynamic amplifiers is throughout explored. A dynamic CMOS differential stage is proposed in [30] and detailed as illustrated in Fig. 2.30, with resource to a single amplification

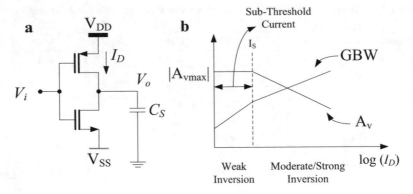

Fig. 2.29 CMOS dependency on current: (**a**) inverter; (**b**) gain and GBW as function of current [30]

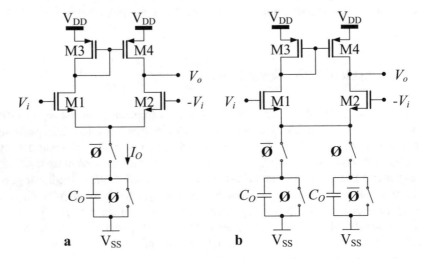

Fig. 2.30 CMOS dynamic behavior: (**a**) differential stage during one phase; (**b**) differential stage during two phases [30]

phase. It is possible to implement the topology in a double-phase configuration, for higher clock rates, as illustrated in Fig. 2.30. Almost 40 years later, the interest in these types of dynamic amplifiers has ramped up, mainly due the low current consumption requirements due to battery lifetime constraints. Some illustrative examples are the CMOS clock and data recovery recently proposed in [32] and the fully dynamic pipeline ADC proposed in [33]. Particularly considering low-power, moderate-resolution, and high-speed ADCs, as well as high-performance SC filters, dynamic amplifiers have special interest. In the work proposed in [32], a dynamic amplification stage is employed as charge-steering logic, specifically to improve the power consumption of the architecture, relying on the possibility of operating with moderate input and output data swings, while drawing power for only a fraction of the duty cycle [32].

In the work presented in [33], a residue amplifier is employed in a dynamic folding stage, to drive a digital output encoder with enhanced speed and improved power consumption, demonstrating the interest of dynamic behavior in state-of-the-art analog-to-digital conversion architectures. In fact, the speed of ADCs is normally limited by the OTAs, which are used to generate and amplify the residues as addressed in the works proposed in [33, 34]. The necessary amount of OTA settling depends on the desired signal-to-noise-and-distortion ratio. However, fast-settling OTAs often consume significant power, which needs to be avoided, especially in the field of wireless and/or portable applications and hardware. In addition to limiting the speed, the OTAs usually govern the power consumption of the complete circuit.

Researchers have been made to either remove OTAs from pipelined A/D converters, mostly motivated by degraded performance in scaled technologies or their high-power consumption [35], or, as found in recent literature [34], reuse the comparator block to perform residue amplification in the context of hybrid implementations, i.e., pipelined SAR-assisted ADC architectures. Therefore, there is an evident need for low-power and high-speed amplifiers to be implemented in low-power, high-speed, and moderate-resolution ADCs and also high-performance SC filters, while being capable of overcoming the challenges of analog design in deeper nanoscale technology nodes and, furthermore, the tendency of using lower voltage supplies. The development and proposal of innovative architectures of dynamic amplifiers with optimized performance is, therefore, entirely justified.

2.8 Summary

A summary of the most relevant state-of-the-art works in the field of single-stage amplifiers is presented and summarized in Table 2.4, with emphasis on the low-frequency gain, evaluated in decibel, GBW, current consumption, and FOM evaluated in MHz × pF/mA and defined in (2.14), in which the tradeoffs involved are highlighted. In this work, the comparison with the state of the art regards only single-stage topologies for fairness of comparison, enforced by the fact that the proposed approaches can be always embedded in multistage amplification architectures, for enhanced performance. This demonstrates the effort of the proposed work, i.e., the development of single-stage amplifiers in terms of gain enhancement strategy and energy-efficient optimization.

Table 2.4 Summary of state-of-the-art single-stage amplifiers in recent literature

Work	Gain [dB]	I_{DD} [mA]	GBW [MHz]	FOM [MHz × pF/mA]	Tech [nm]	Principles and techniques – year	Available OS (Best case)
[1]	60.9	0.800	134.2	939.4	180	Recycling folded cascode (feed-forward)	$OS \approx V_{DD} - 4 \times VDS_{SAT}$
[16]	54.0	1.000	965.0	965.0	180	Recycling folded cascode	$OS \approx V_{DD} - 4 \times VDS_{SAT}$
[17]	85.6	1.000	987.0	987.0	180	Recycling folded cascode	$OS \approx V_{DD} - 4 \times VDS_{SAT}$
[18]	72.0	0.652	159.0	536.5	180	Folded cascode (body driven)	$OS \approx V_{DD} - 4 \times VDS_{SAT}$
[19]	84.0	0.085	12.5	146.0	180	Negative resistance	$OS \approx V_{DD} - 4 \times VDS_{SAT}$
[20]	94.9	3.670	414.0	225.0	130	Folded cascode (negative feedback)	$OS \approx V_{DD} - 4 \times VDS_{SAT}$
[24]	91.5	5.000	714.5	1071.5	130	Folded-cascode (negative feedback)	$OS \approx V_{DD} - 4 \times VDS_{SAT}$
[28]	67.0	2.170	920.0	211.0	180	Folded cascode (positive feedback)	$OS \approx V_{DD} - 4 \times VDS_{SAT}$

References

1. R. Assaad, J. Silva-Martinez, "The Recycling Folded-cascode: A General Enhancement of the Folded-Cascode Amplifier," in IEEE Journal of Solid-State Circuits, Vol. 44, Issue 9, Page(s): 2535–2542, Sep. 2009. DOI: https://doi.org/10.1109/JSSC.2009.2024819.
2. M. Amourah, R. Geiger, "All Digital Transistors High Gain Operational Amplifier Using Positive-feedback Technique," in IEEE International Symposium on Circuits and Systems (ISCAS), Page(s): 701–704, May 2002. DOI: https://doi.org/10.1109/ISCAS.2002.1009937.
3. J. Ko, et al., "D-Band Common-Base Amplifiers With Gain-boosting and Inter stage Self-Matching in 0.18-μm SiGe HBT Technology," in IEEE Transactions on Circuits and Systems-II: Express Briefs, Vol. 64, Issue 3, Page(s): 254–258, Mar. 2017. DOI: https://doi.org/10.1109/TCSII.2016.2561963.
4. Y. Zheng, C. Saavedra, "Feedforward-Regulated Cascode OTA for Gigahertz Applications," in IEEE Transactions on Circuits and Systems-I: Regular Papers, Vol. 55, Issue 11, Pages(s): 3373–3382, Dec. 2008. DOI: https://doi.org/10.1109/TCSI.2008.2001800.
5. M. Ahmadi, "A New Modeling and Optimization of Gain-Boosted Cascode Amplifier for High-Speed and Low-Voltage Applications," in IEEE Transactions on Circuits and Systems-II: Express Briefs, Vol. 53, Issue 3, Page(s): 169–173, Mar. 2006. DOI: https://doi.org/10.1109/TCSII.2005.858493.
6. J. Huijsing, "Operational Amplifiers: Theory and Design," Second Edition, Springer, 2011. DOI: https://doi.org/10.1007/97894-007-0596-8.
7. T. Carusone, et al., "Analog Integrated Circuit Design," Second Edition, Wiley, 2012. ISBN: 978-0470770108.
8. P. Allen, D. Holberg, "CMOS Analog Circuit Design," Second Edition, Oxford, 2002. ISBN: 9780195116441.

9. B. Razavi, "Design of Analog CMOS Integrated Circuits," McGraw-Hill, 2000. ISBN: 978-0072380323.
10. H. Uhrmann, et al., "Analog Filters in Nanometer CMOS," Springer, 2014. ISBN: 978-3-642-38013-6.
11. R. Baker, "CMOS Circuit Design, Layout and Simulation," Third Edition, Wiley, 2010. ISBN: 978-0-470-88132-3.
12. F. Maloberti, "Analog Design for CMOS VLSI Systems," Springer, 2001. ISBN: 978-0-306-47952-6.
13. W. Sansen, "Analog Design Essentials," Springer, 2006. ISBN: 978-0-387-25747-1.
14. P. Gray, et al., "Analysis and Design of Analog Integrated Circuits," Fifth Edition, Wiley, 2009. ISBN: 9781118313091.
15. A. Sedra, K. Smith, "Microelectronic Circuits," Sixth Edition, Oxford, 2009. ISBN: 978-0195323030.
16. K. Bult, G. Geelen, "A fast-settling CMOS op amp with 90 dB DC-gain and 116 MHz unity-gain frequency," in IEEE International Solid-State Circuits Conference (ISSCC), Page(s): 108–109, Feb. 1990. DOI: https://doi.org/10.1109/ISSCC.1990.110152.
17. S. Zhang, et al., "Design of A Low-power, High Speed Op-amp for 10bit 300Msps Parallel Pipeline ADCs," in International Conference on Integration Technology, Page(s): 504–507, Mar. 2007. DOI: https://doi.org/10.1109/ICITECHNOLOGY.2007.4290367.
18. L. Su, Y. Qui, "Design of a Fully Differential Gain-Boosted Folded-Cascode Op Amp with Settling Performance Optimization," in IEEE Conference on Electron Devices and Solid-State Circuits, Page(s): 441–444, Dec. 2005. DOI: https://doi.org/10.1109/EDSSC.2005.1635302.
19. Y. Li, et al., "Transconductance enhancement method for operational transconductance amplifiers," in Electronics Letters, Vol. 46, Issue 19, Page(s): 1321–1323, Sep. 2010. DOI: https://doi.org/10.1049/el.2010.1575.
20. M. Ahmed, et al., "An Improved Recycling Folded-cascode Amplifier with Gain-boosting and Phase Margin Enhancement," in IEEE International Symposium on Integrated Circuits (ISCAS), Page(s): 2473–2476, May 2015. DOI: https://doi.org/10.1109/ISCAS.2015.7169186.
21. S. Zabihian, R. Lofti, "Ultra-Low-Voltage, Low-Power, High-Speed Operational Amplifiers Using Body-Driven Gain-Boosting Technique," in IEEE International Symposium on Circuits and Systems (ISCAS), Page(s): 705–708, May 2007. DOI: https://doi.org/10.1109/ISCAS.2007.377906.
22. M. Fallah, H. Naimi, "A Novel Low Voltage, Low Power and High Gain Operational Amplifier Using Negative Resistance and Self Cascode Transistors," in International Journal of Engineering Transactions C: Aspects, Vol. 26, Issue 3, Page(s): 303–308, 2013.
23. S. Enche, et al., "A CMOS Single-Stage Fully Differential Folded-cascode Amplifier Employing Gain-boosting Technique," in IEEE International Symposium on Integrated Circuits (ISIC), Page(s): 234–237, Dec. 2011. DOI: https://doi.org/10.1109/ISICir.2011.6131939.
24. X. Liu, J. McDonald, "Design of Single-Stage Folded-Cascode Gain-boost Amplifier for 14bit 12.5Ms/S Pipelined Analog-to-Digital Converter," in IEEE International Conference on Software Engineering, Page(s): 622–626, Sep. 2012. DOI: https://doi.org/10.1109/SMElec.2012.6417222.
25. P. Yu, H. Lee, "A High-Swing 2-V CMOS Operational Amplifier with Replica-Amp Gain Enhancement," in IEEE Journal of Solid-State Circuits, Vol. 28, Issue 12, Page(s): 1265–1272, Dec. 1993. DOI: https://doi.org/10.1109/4.261993.
26. E. Sackinger, W. Guggenbuhl, "A High-Swing High-Impedance MOS Cascode Circuit," in IEEE Journal of Solid-State Circuits, Vol. 25, Issue 1, Page(s): 289–298, Feb. 1990. DOI: https://doi.org/10.1109/4.50316.
27. K. Bult, G. Geelen, "A fast-settling CMOS Op Amp for SC Circuits with 90-dB DC Gain," in IEEE Journal of Solid-State Circuits, Vol. 25, Issue 6, Page(s): 1379–1384, Dec. 1990. DOI: https://doi.org/10.1109/4.62165.
28. B. Alizadeh, A. Dadashi, "An Enhanced Folded-cascode Op Amp in 0.18 μm CMOS Process with 67dB Dc Gain," in IEEE Faible Tension Faible Consommation (FTFC), Page(s): 87–90, May 2011. DOI: https://doi.org/10.1109/FTFC.2011.5948926.

29. M. Copeland, J. Rabaey, "Dynamic Amplifiers for MOS Technology," in Electronics Letters, Vol. 15, Page(s): 301–302, May 1979. DOI: https://doi.org/10.1049/el:19790214.
30. B. Hosticka, "Dynamic CMOS Amplifiers," in IEEE Journal of Solid-State Circuits, Vol. 15, Issue 5, Page(s): 887–894, Oct. 1980. DOI: https://doi.org/10.1109/JSSC.1980.1051488.
31. B. Hosticka, et al., "Performance of Integrated Dynamic MOS Amplifiers," in Electronics Letters, Vol. 17, Issue 8, Page(s): 298–300, Apr. 1981. DOI: https://doi.org/10.1049/el:19810209.
32. J. Jung, B. Razavi, "A 25-Gb/s 5-mW CMOS CDR/Deserializer," in IEEE Journal of Solid-State Circuits, Vol. 48, Issue 3, Page(s): 684–697, Mar. 2013. DOI: https://doi.org/10.1109/JSSC.2013.2237692.
33. B. Verbruggen, et al., "A 2.6 mW 6 bit 2.2 GS/s Fully Dynamic Pipeline ADC in 40 nm Digital CMOS," in IEEE Journal of Solid-State Circuits, Vol. 45, Issue 10, Page(s): 2080–2090, Oct. 2010. DOI: https://doi.org/10.1109/JSSC.2010.2061611.
34. M. Gandara, et al., "A Pipelined SAR ADC Reusing the Comparator as Residue Amplifier," in IEEE Custom Integrated Circuits Conference (CICC), Page(s): 1–4, Apr. 2017. DOI: https://doi.org/10.1109/CICC.2017.7993696.
35. M. Anthony, et al., "A process-scalable low-power charge-domain 13-bit pipeline ADC," in IEEE Symposium on VLSI Circuits, Page(s): 222–223, Jun. 2008. DOI: https://doi.org/10.1109/VLSIC.2008.4586015.

Chapter 3
Proposed Family of CMOS Amplifiers

3.1 Voltage-Combiner Structure

Voltage-combiners are typically used in RF circuitry, for converting fully differential signals into a single-ended one, for 50 and 75 Ohm impedance matching of measurement equipment. In this work, however, this configuration is explored as small-signal voltage gain provider, since, as described further in this section, a voltage-combiner (VC) can operate effectively as an amplifier. The electrical scheme of a VC is shown in Fig. 3.1.

A VC employs a combination of an NMOS in common drain accomplished by the device N0 and an NMOS in common source accomplished by the device N1. When not considering the body effect of N0, the small-signal equivalent can be approximated to the circuit presented in Fig. 3.2; and through simple circuit analysis, i.e., applying the Kirchhoff's current law at the output node, it is possible to extract the gain expression of the circuit, as demonstrated through (3.1). Moreover, if the body effect of the common-drain device is considered, as shown in Fig. 3.3, the same type of analysis can be applied resulting, in this particular case, in (3.2). The transfer function (TF) of the circuit presented in Fig. 3.1 can be extracted by directly applying the Kirchhoff laws or, as alternative, using a symbolic analyzer. In this work, the TF is extracted using SapWin 4.0 [1, 2] and is presented in (3.3), where $gds_{N0 + N1} = gds_{N0} + gds_{N1}$ and $cdb_{N0 + N1} = cdb_{N0} + cdb_{N1}$. Through the observation of (3.3), it is possible to verify the existence of the Miller effect through parasitic capacitances cgd_{N1} and cgs_{N0}, the existence of poles and zeros in a VC, and, finally, the order of the transfer function of this circuit, which is one in this case.

$$gm_{N0}(v_i - v_o) = gds_{N0}(v_o) + gds_{N1}(v_o) + gm_{N1}(-v_i) \qquad (3.1)$$

$$gm_{N0}(v_i - v_o) = gds_{N0}(v_o) + gds_{N1}(v_o) + gm_{N1}(-v_i) + gmb_{N0}(v_{sb} = v_o) \qquad (3.2)$$

© Springer International Publishing AG, part of Springer Nature 2019
R. F. S. Póvoa et al., *A New Family of CMOS Cascode-Free Amplifiers with High Energy-Efficiency and Improved Gain*,
https://doi.org/10.1007/978-3-319-95207-9_3

Fig. 3.1 Single-ended
voltage-combiner

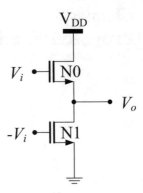

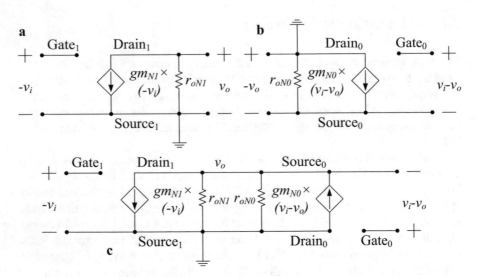

Fig. 3.2 VC small-signal equivalent: (**a**) common-source; (**b**) common-drain; (**c**) complete circuit

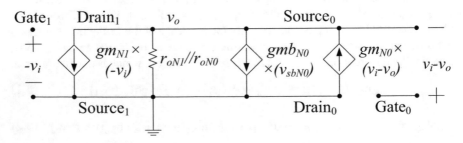

Fig. 3.3 VC small-signal equivalent with body effect

$$\text{TF}_{\text{VC}}(s) = \frac{gm_{\text{N1}} + gm_{\text{N0}} + (cgs_{\text{N0}} - cgd_{\text{N1}}) \times s}{gm_{\text{N0}} + gds_{\text{N0+N1}} + (cdb_{\text{N0+N1}} + cgs_{\text{N0}} + cgd_{\text{N1}}) \times s} \tag{3.3}$$

For simplicity of analysis and in order to allow some circuit insight, minor simplifications are used in the derived equations, namely, the body effect of the transistors is neglected. From the transfer function, it is fairly easy to obtain the low-frequency open-circuit gain, i.e., the low-frequency gain of the VC circuit, e.g., it is enough to calculate the limit of (3.3) when s tends to zero. The open-circuit gain of the circuit is presented in (3.4). Considering gm_{N0} and $gm_{\text{N1}} \gg gds_{\text{N0 + N1}}$, a good approximation can be given by (3.5). By sizing the circuit to have $gm_{\text{N0}} \approx gm_{\text{N1}}$, a voltage gain of 6 dB can be effortlessly achieved. However, this value of gain is not limitative, since there is a degree of freedom in the sizing of the transconductance of both active devices. Considering modern technology, e. g., the UMC 130 nm CMOS technology, it is possible to achieve a gain of approximately 10 dB with the proposed topology, without penalty in dynamic range and with proper DC biasing. In this work, the potential of VC circuits and corresponding implementation variants is explored in terms of additional voltage gain in the designed OpAmps, in the sense that the gain of the VCs is added to the intrinsic gain of the amplifier, leading to higher GBW values, without significant penalty in terms of power consumption, i.e., higher energy-efficient values. The factor of one in the gain expression is mainly due to the contribution of the common-drain device, i.e., N0, while the common-source device, i.e., N1, is important to maximize the second term. Furthermore, the existing single high-frequency pole, ω_{P}, and the gain-bandwidth product can be given by (3.6) and (3.7), respectively.

Following the sizing strategy depicted in Fig. 3.4, a single-ended VC structure similar to the one presented in Fig. 3.1 is implemented using the UMC 130 nm technology node, using standard-VT NMOS devices, and sized with $gm_{\text{N0}} \approx 50$ mΩ^{-1} and $gm_{\text{N1}} \approx 60$ mΩ^{-1}, considering a 6 pF load. The dimensions of the devices and the basic DC parameters of the devices are presented in Table 3.1 and in Table 3.2, correspondingly. Both of the devices are well saturated, and the current consumption, indicated in Table 3.2, refers to the complete circuit consumption. These results correspond to an implementation in which transistor N0 does not suffer of body effect, i.e., the source voltage is equal to the bulk voltage. In fact, the

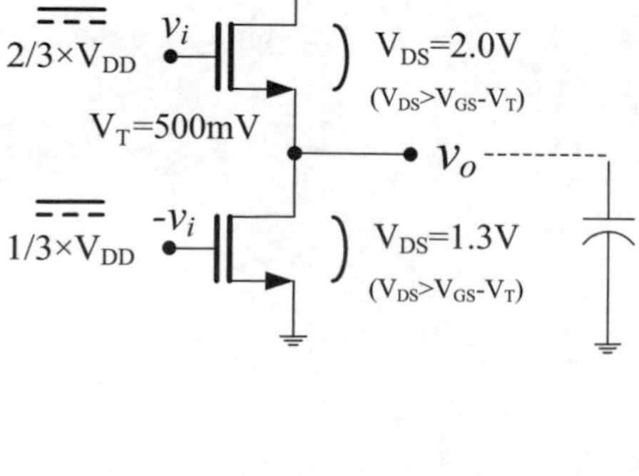

Fig. 3.4 Single-ended voltage-combiner generic sizing strategy and test bench

Table 3.1 VC device sizing

Device parameter	Dimension
l_{N0}	520 nm
w_{N0}	27.44 μm
nf_{N0}	14
l_{N1}	1.78 μm
w_{N1}	87.06 μm
nf_{N1}	4

Table 3.2 Simulated DC parameters of the VC

Circuit parameter	Value
V_{DD}	3.3 V
VCM_{N0}	2.3 V
VCM_{N1}	1.0 V
VGS_{N0}	1.0 V
VGS_{N1}	1.0 V
$VGS_{N0}\text{-}VT_{N0}$	490 mV
$VGS_{N1}\text{-}VT_{N1}$	550 mV
VDS_{N0}	2.0 V
VDS_{N1}	1.3 V
$VDS_{N0}\text{-}VDS_{SATN0}$	1.6 V
$VDS_{N1}\text{-}VDS_{SATN1}$	800 mV

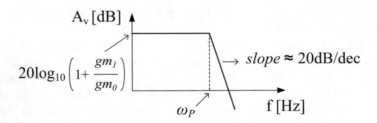

Fig. 3.5 Single-ended voltage-combiner asymptotic frequency response

small-signal response of the VC in the frequency domain respects a low-pass filter with the single high-frequency pole given as in (3.6); thus, it is expected a response that follows the simplified single-pole frequency response, in which the low-frequency gain is given as in (3.5), as shown in Figs. 3.4 and 3.5.

$$A_{vvc} = (gm_{N1} + gm_{N0})/(gm_{N0} + gds_{N0+N1}) \tag{3.4}$$

$$A_{vvc} \approx \left(1 + \frac{gm_{N1}}{gm_{N0}}\right) \tag{3.5}$$

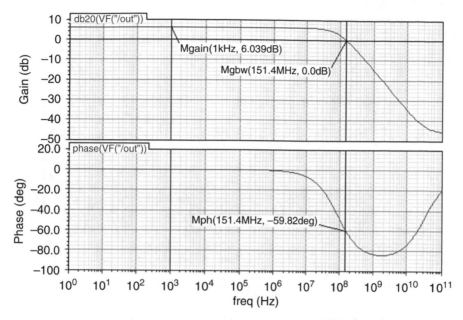

Fig. 3.6 Simulated single-ended voltage-combiner small-signal response

$$\omega_P = \frac{gm_{N0} + gds_{N0+N1}}{cdb_{N0+N1} + cgs_{N0} + cgd_{N1}} \tag{3.6}$$

$$GBW_{VC} = \frac{gm_{N1} + gm_{N0}}{cdb_{N0+N1} + cgs_{N0} + cgd_{N1}} \tag{3.7}$$

The AC simulation results of a single-ended VC, described previously, are shown in Fig. 3.6, depicting a first-order low-pass filter frequency response, with a low-frequency gain higher than 0 dB. This means this structure represents a non-inverting active amplification scheme. Through Fig. 3.7, it is possible to observe a single pole and zero, depicting a single pole located in a frequency of approximately 87.2 MHz. The structure provides a low-frequency gain of approximately 6 dB, and a considerable low noise contribution, as illustrated in the equivalent input noise response in Fig. 3.8, depicting an approximate value of 3.3 nV/√Hz at the value of GBW, which in this case is approximately equal to 151.4 MHz. The performance metrics of the circuit are summarized in Table 3.3.

The following paragraph addresses a 3-σ Monte Carlo simulation with 100 runs carried out using Cadence® framework. Briefly overlooking elementary statistical analysis, the 3-σ represents the standard deviation, i.e., the metric that is used to quantify the amount of variation of a set of data values [3]. In normal distributions, e.g., Monte Carlo tryouts, a 3-σ standard deviation corresponds to a 99.7% confidence interval, which means, literally, that approximately 100 out of 100 samples behave as the results shown in Fig. 3.9 suggest. Maximum deviations of approximately 2.9% and 3% are achieved, in the low-frequency gain and in the input-

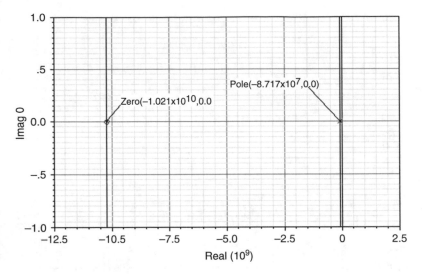

Fig. 3.7 Simulated single-ended voltage-combiner pole-zero locations: × poles, ○ zeros

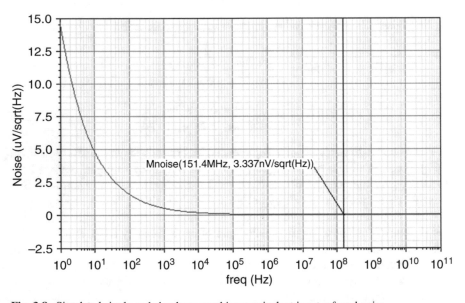

Fig. 3.8 Simulated single-ended voltage-combiner equivalent input-referred noise

referred noise at GBW, correspondingly. The gain improvement of the simulated VC structure is, in the conditions detailed in the previous paragraphs, robust to process and mismatch variations.

In this section, a single-ended VC structure is addressed, both at analytical and simulation level. The single-ended VC corresponds to two NMOS devices, in common-gate and common-source configuration which, with proper sizing, are

Table 3.3 Simulated key performance parameters of the sized VC structure

Metrics	Performance
GBW	151.4 MHz
Gain	6.0 dB
I_{DD}	1 mA
Noise	3.34 nV/√Hz
Phase at GBW	59.8°

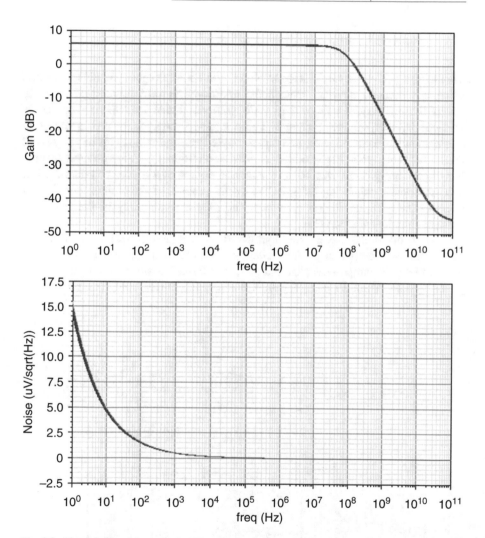

Fig. 3.9 Simulated 3-σ Monte Carlo 100 runs of a single-ended VC equivalent input-referred noise

able to provide active gain. In particular, the frequency response of the VC structure follows the response of an active low-pass filter, with no phase inversion. The structure demonstrates robustness through Monte Carlo simulations with process and mismatch. Through the course of the present work, the direct advantages of replacing a typical static current biasing source, by a VC-based structure, with proper modifications to sustain proper DC biasing for lower-supply voltages, in order to enhance the gain and GBW of an OTA, leading to high values of energy efficiency, when compared to the state of the art in the field of single-stage amplifiers, are demonstrated.

3.2 Voltage-Combiner-Biased OTA

This section presents a single-stage amplifier with enhanced gain without the need of using any cascode devices. Instead, the traditional current source that typically biases the differential pair of a given operational transconductance amplifier is replaced by two voltage-combiners, i.e., a structure comprising a common-drain and a common-source device, in a cross-coupled configuration. This has a twofold effect: first, the two voltage-combiners provide additional gain themselves; second, the differential-pair devices act as a common-source and a common-gate device, simultaneously, enhancing the gain-bandwidth product of the circuit, therefore its energy efficiency, as described further. The proposed topology is, therefore, the voltage-combiner-biased OTA.

The electrical scheme of the amplifier presented in this section is shown in Fig. 3.10, with emphasis on the voltage-combiners. This amplifier comprises a differential pair composed by the N0 pair, two PMOS basic current mirrors composed by both P0 and P1 pairs, and an NMOS current mirror, composed by N3 pair.

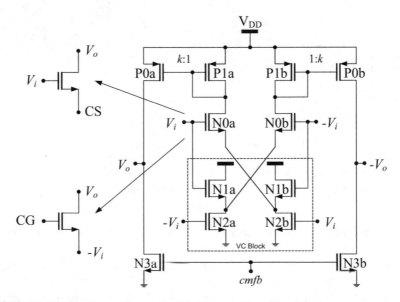

Fig. 3.10 Electrical scheme of the proposed voltage-combiner-biased OTA

The innovation proposed in this circuit, when compared to the classic symmetrical CMOS OTA, relies on the fact that the traditional NMOS current source that biases the differential pair is in fact removed. Instead, two voltage-combiners in a cross-coupled configuration are used. The first VC is represented by transistors N1a and N2a, while the second VC is represented by transistors N1b and N2b. Moreover, it is possible to verify, with the help of the electrical scheme shown in Fig. 3.10, that both devices of the differential pair, i.e., N0a and N0b, implement at the same time the function of common source, i.e., input signal, V_i or $-V_i$, and output current, I, and the function of common gate, i.e., input signal, $-A_{v1} \times v_i$ or $A_{v2} \times -v_i$, and output current, I. Therefore, a single device, either N0a or N0b, implements, by itself, a single-device cascode structure.

The complete gain expression of the amplifier is rather complex to extract with classic small-signal circuit analyses. However, with recourse to SapWin 4.0 symbolic analyzer [2], the expression can be easily extracted and is presented in (3.8). The output linear conductance of a single-ended VC structure is given by (3.10). For simplification of analysis, the body effect of the devices N0 and N1 is neglected in the following expressions. The gain of the amplifier is, by (3.8), proportional to the transconductance of the differential-pair elements composed by the N0 pair; however, this metric is boosted with a proportion of the transconductance of the N2 pair. Moreover, the gain is also inversely proportional to the transconductance of the N1 pair, relating straight to the approximation in (3.5), where the gain expression of the VC structure is shown. A generic design strategy is depicted in Fig. 3.11.

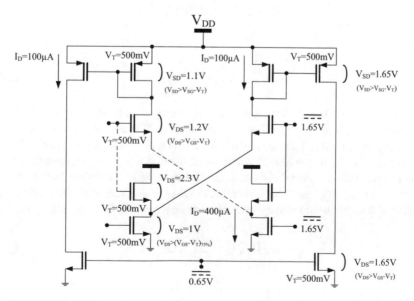

Fig. 3.11 VC-biased OTA sizing strategy

$$A_{\text{VOTA}} = \frac{gm_{\text{N1}}(gds_{\text{N0}} + 2gm_{\text{N0}}) + gm_{\text{N2}}(gds_{\text{N0}} + gm_{\text{N0}}) + gm_{\text{N0}}gds_{\text{VC}}}{\begin{bmatrix} gm_{\text{P1}}(gds_{\text{N0}} + gm_{\text{N0}} + gm_{\text{N1}} + gds_{\text{VC}}) + gds_{\text{N0}}gds_{\text{VC}} + \\ gm_{\text{N1}}(gds_{\text{N0}} + gds_{\text{P1}}) + gds_{\text{P1}}(gm_{\text{N0}} + gds_{\text{N0}} + gds_{\text{VC}}) \end{bmatrix}}$$

$$\times \frac{gm_{\text{P0}}}{gds_{\text{O}}} \tag{3.8}$$

where

$$gds_{\text{O}} = gds_{\text{P0}} + gds_{\text{N3}} \tag{3.9}$$

and

$$gds_{\text{VC}} = gds_{\text{N1}} + gds_{\text{N2}} \tag{3.10}$$

The active devices are designed to operate in saturation, where it is important to consider the overdrive voltage, i.e., $V_{\text{GS}} - V_{\text{T}}$, delivering the value of current flowing through the device as given by the quadratic model shown in (3.11). For the UMC 130 nm technology node, the device models show roughly the following: $\mu_{\text{n}} \approx 3 \times 10^{-2}$ m^2/Vs, $C_{\text{oxn}} \approx 5 \times 10^{-3}$ F/m^2, $t_{\text{oxn}} \approx 7 \times 10^{-9}$ m, $\mu_{\text{p}} \approx 9 \times 10^{-3}$ m^2/Vs, $C_{\text{oxp}} \approx 5 \times 10^{-3}$ F/m^2, $t_{\text{oxp}} \approx 7 \times 10^{-9}$ m, and $\varepsilon_{\text{oxn}} = 35 \times 10^{-12}$ F/m. Through simulation, for the p-type MOSFETs, a $\mu_{\text{p}}C_{\text{ox}} \approx 30$ μAV^{-2} is simulated, whereas the n-type MOSFETs shown a $\mu_{\text{n}}C_{\text{ox}} \approx 200$ μAV^{-2}. It is relevant to keep in mind that the mobility is proportional to the carrier relaxation time and inversely proportional to the carrier effective mass; hence, the electrons have generically a higher mobility than the holes.

$$I_{\text{D}} \approx \frac{1}{2}\mu C_{\text{ox}}\frac{W}{L}(V_{\text{GS}} - V_{\text{T}})^2 \tag{3.11}$$

where

$$C_{\text{ox}} = \frac{\varepsilon_{\text{ox}}}{t_{\text{ox}}} \tag{3.12}$$

Assuming that the devices are to be operating in saturation, the condition to consider is $V_{\text{DS}} > V_{\text{GS}} - V_{\text{T}}$. However, the devices N2a and N2b can be sized with small channel lengths. This has a twofold effect: on one hand, the gm of the devices, as given by (3.13), is maximized; on the other hand, this property enables the devices to suffer from short-channel effects, in particular, velocity saturation. In this case, the drain-source voltage has to be higher than VDS$_{\text{SAT}}$, as given by (3.14), where κ is a metric of the degree of velocity saturation, as described in the following paragraph.

$$gm = \sqrt{2\mu C_{\text{ox}}\left(\frac{W}{L}\right)I_{\text{D}}} = \frac{2I_{\text{D}}}{V_{\text{GS}} - V_{\text{T}}} \tag{3.13}$$

Fig. 3.12 Velocity-saturation effect

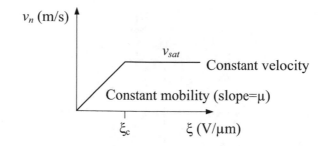

$$\text{VDS}_{\text{SAT}} = \kappa(V_{\text{GS}} - V_{\text{T}}) \times (V_{\text{GS}} - V_{\text{T}}) \quad (3.14)$$

In fact, VDS_{SAT} becomes lower than $V_{\text{GS}} - V_{\text{T}}$, and, therefore, the saturation condition is easily met in current short-channel devices [4]. Through experimental procedures based on simulation, VDS_{SAT} is, in general, approximately equal to 75% of the overdrive voltage, if the devices of UMC 130 nm technology node are considered. The behavior of transistors with very short-channel lengths, i.e., short-channel devices, deviates from the principle of carrier mobility consistency with weaker field strengths, when compared to regular sizing. In physical terms, when the electrical field along the channel of a device reaches the threshold value of ξ_c, the velocity of the carriers tends to saturate due to scattering effects, i.e., collisions suffered by the carriers, which happens earlier, if short-channel devices are considered. The velocity-saturation effect is illustrated in Fig. 3.12. In this approach, κ is a metric of the degree of velocity saturation and is given by eq. (3.15). As V_{DS} approached VDS_{SAT}, the drain current can be approximately given by (3.16).

The following analyses consider an open-circuit simulation test bench, as illustrated in Fig. 3.13, allowing the direct extraction at simulation level [5] of the low-frequency gain, GBW, phase margin, and offset voltage. The nominal capacitive load at the output of the circuit is 6 pF, which derives from the expected value of a two-layer printed circuit board [7], targeting experimental verification described further in this work. Note that an ideal and continuous-time common-mode feedback (CMFB) circuit is employed; this is a fully differential implementation. The CMFB circuit implements directly the expression in (3.17), where the differential VOS is compensated in the cmfb voltage node, taking into consideration the V_{bias} voltage that directs the output. Both N0 and N1 devices do not suffer from body effect, i.e., the source voltages equal the bulk voltages, targeting physical implementation in a triple-well technology.

$$\kappa(V) = \frac{1}{1 + \left(\frac{V}{\xi_c \times L}\right)} \quad (3.15)$$

$$I_{\text{DSAT}} = v_{\text{sat}} C_{\text{ox}} W (V_{\text{GS}} - V_{\text{T}} - \text{VDS}_{\text{SAT}}) \quad (3.16)$$

$$\text{cmfb} = -\left(^{V_{\text{DD}}}/_2 - \left(^{V_{\text{out}+}}/_2\right) - \left(^{V_{\text{out}-}}/_2\right) - V_{\text{bias}}\right) \quad (3.17)$$

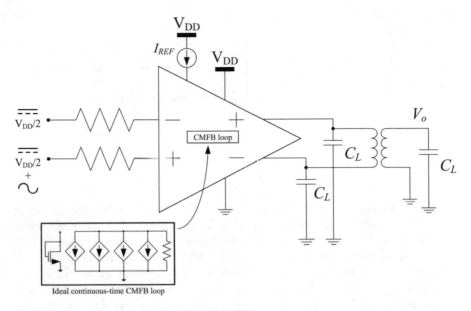

Fig. 3.13 Open-circuit test bench with ideal CMFB

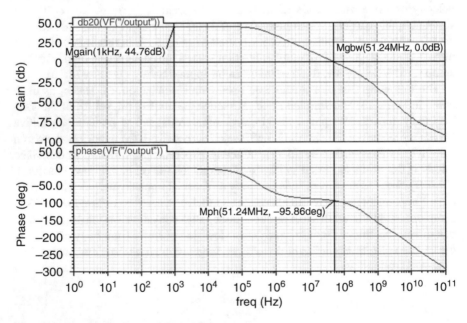

Fig. 3.14 Simulated VC-biased OTA AC response

The AC response of the OTA is presented in Fig. 3.14, depicting a gain of 45 dB, a GBW of 51 MHz, and a PM higher than 60°, i.e., 84.1°, showing sufficient stability for the applications targeted for a single-stage OTA. Each device of the differential pair drains approximately 115 μA. If instead of the cross-coupled VCs used to bias

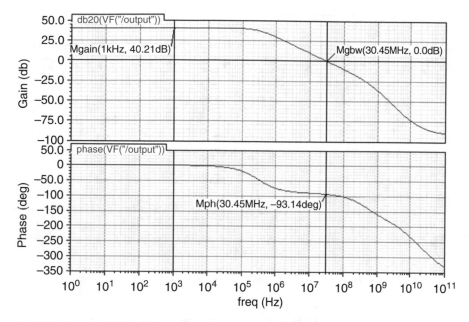

Fig. 3.15 Simulated OTA AC response with conventional biasing

the differential pair, an equivalent current source of 230 µA is employed, the overall gain of the amplifier drops approximately 4.6 dB, and the GBW of the OTA reduces by half, as Fig. 3.15 shows. In both cases, an offset voltage of approximately 9 mV can be obtained. The current consumption of this implementation is approximately equal to 1.3 mA, drained from a 3.3 V voltage supply, not considering I_{REF}. The gain and GBW values prove the additional contribution by biasing the differential pair of the CMOS OTA with two VCs, instead of a static current source.

In this section, a VC-biased OTA is addressed at analytical level, compounded with an implementation at sizing level and simulation results. The proposed amplifier topology derives initially from the symmetrical CMOS OTA. However, this amplifier employs two single-ended VC structures in a cross-coupled configuration, in replacement of the traditional static current source that biases the differential pair. The proposed innovation provides an enhancement of gain and GBW, when compared to a static current biasing approach, i.e., an improvement in the energy efficiency of the original OTA topology.

3.3   Voltage-Combiner-Biased OTA with Current Starving

This section addresses the growing need for energy efficiency, by proposing an upgrade to a voltage-combiner-biased single-stage amplifier; enhancing its low-frequency gain; improving gain-bandwidth product of the amplifier, i.e., reducing

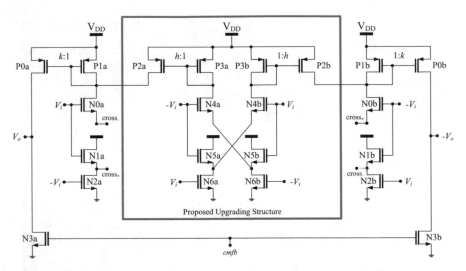

Fig. 3.16 Electrical scheme of the proposed VC-biased OTA with current starving

its settling time; and remarkably improving the power efficiency of the topology described in the previous section. The extension to deeper nanoscale nodes, e.g., 65 or 40 nm, and lower-supply voltages, e.g., 2 V, is relatively straightforward. For supply voltages of 1.2 V or below, the voltage-combiners need to be implemented in a folded configuration fashion to be properly biased in DC, as referred in the topology described in the previous section.

The conventional current starving technique makes use of current that otherwise would sink into ground, through a parallel path, to improve the gain of the amplifier [6]. In the proposed approach, a part of the current used to feed the first differential pair, composed by N0a and N0b, is taken away by current sources and used to feed a second differential pair composed by N4a and N4b, also biased by voltage-combiners, as presented in Fig. 3.16. It is important to keep in mind that the current gain in this topology, which will be reflected in the overall gain of the amplifier, depends only on the mirroring factor and not on the absolute value of current flowing in the respective branches. As a result of less current flowing through the diode connected P1a and P1b, the transconductance of these devices reduces, improving by itself the overall gain of the amplifier, as visible through the gain equation, as is presented, later on, in this section. The sizing of the two pairs of VCs follows $gm_{N2} > gm_{N1}$ and $gm_{N6} > gm_{N5}$.

When compared to the conventional VC-biased CMOS OTA, the proposed structure improves the GBW and settling time and enhances the gain of the amplifier with no significant power dissipation penalty, as experimental results will show. Moreover, the voltage swing is not affected at all since the added structure is in parallel with the existing one; from the small-signal point of view, i.e., no cascode structures are used. In particular, the proposed topology exhibits also a first-order low-pass filter-shaped frequency response. The low-frequency gain expression of the amplifier is presented in (3.18) following a small-signal circuit analysis and making use of SapWin 4.0 symbolic analyzer to extract the expression [2]. The body effect of the devices is neglected. If the gds of the devices in the VCs is negligible when comparing to ones in the differential pairs, (3.19) simplifies into (3.21), which is, as quoted before, maximized when the gm of active loads P3a and P3b is reduced.

$$A_{\text{VOTA}} = A_{\text{VINSIDE\_BRANCHES}} \times A_{\text{VOUTSIDE\_BRANCHES}} \tag{3.18}$$

where

$$A_{\text{VINSIDE\_BRANCHES}} \approx \frac{gm_{P2}gm_{N4}\left(\begin{array}{l}(gds_{N5} + gds_{N6} + gm_{N6}) \times (gds_{N0} + gds_{N1} + gds_{N2}) \\ +gm_{N1}(gds_{N5} + gds_{N6})\end{array}\right)}{gm_{P3}gm_{P1}\left(\begin{array}{l}(gds_{N4} + gds_{N5} + gds_{N6} + gm_{N4} + gm_{N5}) \times \\ (gds_{N0} + gds_{N1} + gds_{N2}) \\ +(gm_{N0} + gm_{N1}) \times (gds_{N4} + gds_{N5} + gds_{N6})\end{array}\right)} \\ + \frac{gm_{N0}gds_{N4}\left(\begin{array}{l}(gds_{N1} + gds_{N2} + gm_{N5} + gm_{N2}) \times \\ (gds_{N5} + gds_{N6})\end{array}\right)}{\phantom{gm_{P3}gm_{P1}}} \tag{3.19}$$

and

$$A_{\text{VOUTSIDE\_BRANCHES}} = gm_{P0}/gds_{P0+N3} \tag{3.20}$$

$$A_{\text{VINSIDE\_BRANCHES}} \approx \frac{gm_{P2}gm_{N4}gm_{N6}gds_{N0} + gm_{N0}gds_{N4}gm_{N5}gm_{N2}}{gm_{P3}gm_{P1}(gds_{N0}(gm_{N5} + gm_{N4} + gds_{N4}) + gds_{N4}(gm_{N0} + gm_{N1}))} \tag{3.21}$$

The transfer function of the OTA described in this section is shown in (3.22). A generic design strategy is depicted in Fig. 3.17, which allows verifying the performance improvement. Using the open-circuit test bench depicted in Fig. 3.13, it is possible to acquire the low-frequency gain, GBW, and the phase margin, at simulation level. Considering the fully differential configuration presented, an ideal and continuous-time CMFB circuit similar to the one shown in Fig. 3.13 is needed. In this case, the N0, N1, N4, and N5 pairs do not suffer from body effect, i.e., the source voltages equal the substrate voltages. The active devices are sized to operate in saturation, where it is important to consider the overdrive voltage, i.e., $V_{\text{GS}} - V_{\text{T}}$, following the quadratic model shown in (3.11) for the value of current flowing through the device, as in the case addressed in the previous section, i.e., the VC-biased OTA. Similar to the previous case, the V_{T} of the N2 and N6 pairs is set to the minimum value allowed by the technology node, in order to take advantage of the short-channel effects. The AC response of the amplifier is presented in Fig. 3.18, showing a gain of 63 dB, a GBW of roughly 146 MHz, and a PM of approximately 60°, i.e., demonstrating a considerable improvement when compared to the values of the VC-biased OTA: a gain of 45 dB and a GBW of 51 MHz. This implementation drains approximately 2 mA, from a 3.3 V source, not considering I_{REF}.

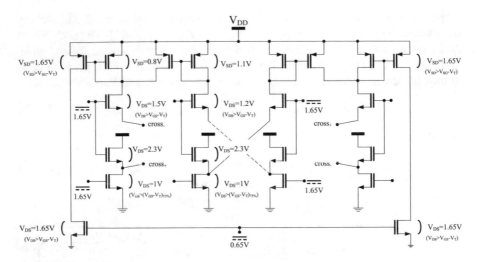

Fig. 3.17 VC-biased OTA with current starving sizing strategy

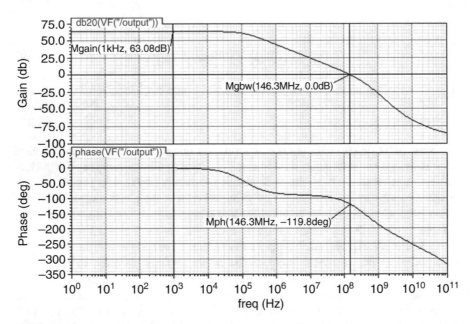

Fig. 3.18 Simulated VC-biased OTA with current starving initial sizing AC response

$$\text{TF} \approx \frac{\text{num}[A_{\text{vOTA}}] + (gm_{P3}gm_{N4}gm_{N1}cgd_{N0})s + (gm_{N5}gm_{N0}cgd_{P2}cgd_{N0})s^2}{\text{den}[A_{\text{vOTA}}] + (gm_{P1}gm_{N4}gm_{N1}cgd_{P2})s + (gm_{P3}gm_{N1}cgs_{N5}cgd_{N0})s^2} \quad (3.22)$$

In this section, a VC-biased OTA with current starving is addressed at analytical level, compounded with an implementation at sizing level and simulation results.

The proposed amplifier topology derives from the VC-biased OTA. However, this amplifier employs two VC-biased structures in a parallel configuration, from the small-signal point of view, and takes advantage of the current starving technique. The proposed innovation provides a significant enhancement of gain and GBW, when compared to a VC-biased OTA, yet, with higher active area and complexity.

3.4 Folded Voltage-Combiner-Biased OTA

This section addresses a response to the tendency of low-voltage supplies, by proposing a topology in which the tail current source that biases the differential pair is replaced by a fully differential folded voltage-combiner block, i.e., a differential input combined with a folded common-drain and a common-source structure, connected in a specific way that has a twofold effect: first, the fully differential folded VC block provides additional low-frequency gain itself; second, the differential-pair devices act, simultaneously, as a common-source and a common-gate device. The extension to deeper nanoscale nodes and lower-supply voltages, e.g., 0.9 V, is relatively straightforward, as described later on, in Chap. 4.

The electrical scheme of a fully differential folded VC is shown in Fig. 3.19. The circuit employs two PMOS common-source devices, i.e., M3a and M3b, and two folded NMOS common-drain transistors, i.e., M2a and M2b, combined with a differential input, while being properly biased by the M1, M4a, and M4b current sources. Following the same type of analysis shown in Section 3.1, it is possible to determine the small-signal equivalent circuit of this topology, as shown in Fig. 3.20, where the body effect of M2 and M3 is not considered, for simplicity of analysis. The low-frequency gain of the fully differential folded VC circuit, A_v, can be approximated to the expression (3.24), neglecting the body effect of M2 and M3, for simplicity of analysis, and after resolving (3.23). Considering that gm_2

Fig. 3.19 Fully differential folded VC structure

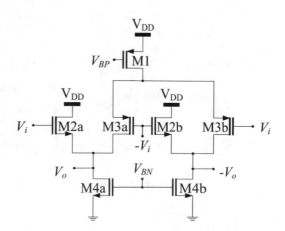

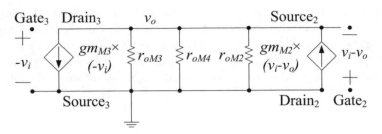

Fig. 3.20 Fully differential folded VC structure small-signal equivalent circuit

and $gm_3 \gg (gds_2 + gds_3 + gds_4)$, a fair approximation of the gain shown in (3.24) is given by (3.25).

$$gm_2(v_i - v_o) = gds_2(v_o) + gds_3(v_o) + gds_4(v_o) + gm_3(-v_i) \tag{3.23}$$

$$A_{vvc} = \frac{gm_2 + gm_3}{gm_2 + gds_2 + gds_3 + gds_4} \tag{3.24}$$

$$A_{vvc} \approx 1 + \frac{gm_3}{gm_2} \Rightarrow |A_{vvc}| > 1 \tag{3.25}$$

As detailed previously, the main advantage of this folded VC circuit can be offered through (3.25), when compared with a basic common-drain, i.e., source-follower circuit, which relies on the fact that the gain can be made higher than two, by properly sizing $gm_3 > gm_2$. The GBW of this VC is given by (3.26). Furthermore, the small-signal output resistance, r_o, of the circuit shown in Fig. 3.19, can be represented by (3.27). Using the proposed VC circuit, the conventional NMOS current source of the differential pair of a traditional single-stage current-mirror OTA can be replaced, yielding the proposed amplifier shown in Fig. 3.21. Following the same type of circuit analysis and after resolving (3.28), the low-frequency gain of the amplifier, A_v, can be given by (3.29), where k represents the mirroring factor and gds_{out} is the output conductance, which is dependent on P0 and N3, as presented in (3.30). Note that no cascode devices are stacked at the output nodes, thus maximizing the output swing of the amplifier. The fully differential folded VC structure requires external biasing current mirrors. Moreover, the gain can also follow (3.8), hence becoming (3.31), with the proper adjustments given by (3.33).

$$GBW = \frac{gm_2 + gm_3}{cdb_3 + cgd_3 + cgs_2 + cgd_4 + cdb_4} \tag{3.26}$$

$$r_o = \frac{1}{gm_2 + gds_3 + gds_4} \cong \frac{1}{gm_2} \tag{3.27}$$

$$-v_o = (k \times gm_{N0} \times (-v_i - (v_i \times A_{vvc})))/gds_{out} \tag{3.28}$$

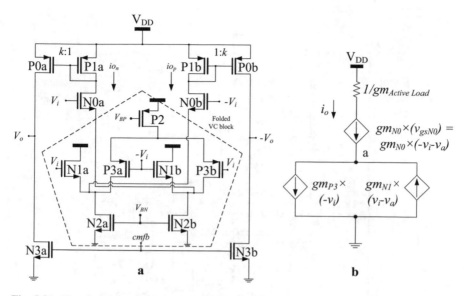

Fig. 3.21 Electrical scheme of the folded VC-biased OTA: (**a**) topology; (**b**) simplified small-signal equivalent

$$A_{VOTA} \approx \frac{k \times io}{gds_{Out}} = \frac{k \times gm_{N0} \times (1 + A_{vvC})}{gds_{Out}} \tag{3.29}$$

$$gds_{Out} = gds_{P0} + gds_{N3} \tag{3.30}$$

$$A_{VOTA} = \frac{gm_{N1}(gds_{N0} + 2gm_{N0}) + gm_{P3}(gds_{N0} + gm_{N0}) + gm_{N0}gds_{VC}}{\left[\begin{array}{c} gm_{P1}(gds_{N0} + gm_{N0} + gm_{N1} + gds_{VC}) + gds_{N0}gds_{VC}+ \\ gm_{N1}(gds_{N0} + gds_{P1}) + gds_{P1}(gm_{N0} + gds_{N0} + gds_{VC}) \end{array}\right]}$$
$$\times \frac{gm_{P0}}{gds_O} \tag{3.31}$$

where

$$gds_O = gds_{P0} + gds_{N3} \tag{3.32}$$

and

$$gds_{VC} = gds_{N1} + gds_{P3} + gds_{N2} \tag{3.33}$$

A generic design strategy is shown in Fig. 3.22. Since the biasing voltage of this configuration is lower, the advantage of short-channel sizing is taken into consideration. However, with smaller channel lengths, the V_T tends to be higher, which is not desirable due to the penalty in dynamic range. In fact, regarding the UMC 130 nm technology node, experimental verification allows concluding generically that the V_T of

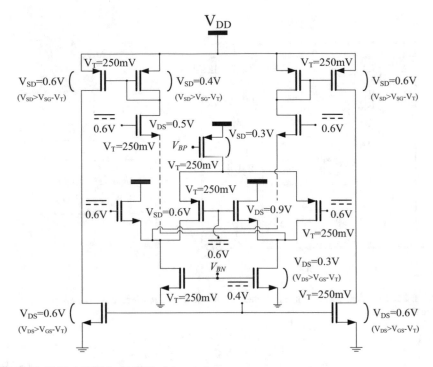

Fig. 3.22 Folded VC-biased OTA sizing strategy

the active devices drops slightly after reaching a peak value, which occurs when the channel length approximately equals twice its minimum value. After surpassing this point, i.e., enhancing the channel length, the V_T drops to asymptotically reach a value approximately equal to 75% of its maximum value, as briefly illustrated in Fig. 3.23. Therefore, all the devices comprised in the voltage-combiners can be sized with channel lengths of 340 nm, i.e., approximately three times the minimum value of length allowed by the technology, hence representing an agreement between low V_T and low VDS_{SAT}, which is particularly important when low biasing voltages are used. With the open-circuit test bench depicted in Fig. 3.13, it is possible to simulate and acquire the low-frequency gain, GBW, offset voltage, and PM. The AC response of the implemented amplifier is presented in Fig. 3.24, depicting a gain of 40.2 dB, a GBW of 67.55 MHz, and a PM of approximately 73°. In this case, the N0, N1, and P3 devices do not suffer from body effect, i.e., the source voltages equal the substrate voltages. This implementation drains approximately 0.5 mA, from a 1.2 V voltage supply source.

In this section, a folded voltage-combiner-biased OTA is addressed at analytical level, compounded with an implementation at sizing level and simulation results. The folded voltage-combiners are used in replacement of a static current source, to bias the differential pair of an OTA. The folded configuration of the VC structure addresses low voltage supplies, avoiding stacking issues and maintaining proper DC biasing. The folded voltage-combiner-biased OTA allows operating with a low systematic offset, with voltage supplies from 1.2 V down to 0.9 V.

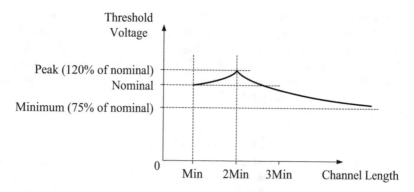

Fig. 3.23 Progression of the threshold voltage with the increase of channel length in a 130 nm technology

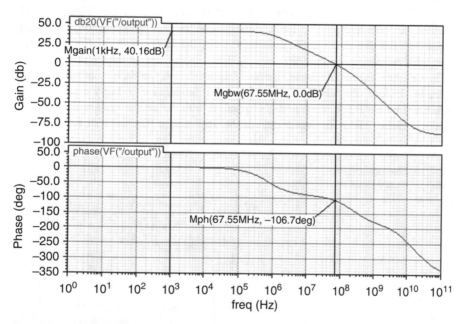

Fig. 3.24 Simulated folded VC-biased OTA AC response

3.5   Dynamic Voltage-Combiner-Biased OTA

This section presents the design of a dynamic voltage-combiner-biased CMOS OTA, for low-power high-speed analog-to-digital converters and high-performance switched-capacitor filters, using standard-VT 1.2 V devices. As mentioned before, the first dynamic OTA is proposed by Copeland and Babey [7] and later explored by Hosticka [8, 9], where the idea behind is to bias the circuit in current during the amplification phase only, and cutting off the current during the reset phase, to improve the power consumption of the circuit. Moreover, since the biasing current achieves its nominal peak value in the early beginning of the amplification phase, the speed of the circuit is, therefore, maximized. Oppositely, since the current is reduced exponentially, the peak low-frequency gain is achieved in the end of this phase, as necessary, by maximizing the gm/I ratio.

The demand for low-power, moderate-resolution, and high-speed ADCs, e.g., as residue amplifier or comparator, as well as high-performance SC filters, requires new dynamic single-stage OTAs, to be readily used either in open-circuit or in closed-circuit. The electrical scheme of the new dynamic proposed OTA is presented in Fig. 3.25. The amplifier described in this section is a fully dynamic implementation of the static approach presented in the previous section of this work. In this circuit, the traditional tail current source that biases the differential pair is replaced by a dynamic fully differential folded VC block, i.e., a differential input combined with a folded common-drain and common-source structure. This has a threefold effect: (1) the VC block provides additional low-frequency gain; (2) the differential-pair devices act, simultaneously, as a common-source and a common-gate device, improving the energy efficiency; and (3) both the GBW and the low-frequency gain are dynamically maximized.

In order to further increase the low-frequency gain without the need of any cascode devices, the two highlighted PMOS transistors in latch configuration, composed by the P2 pair, are added in parallel with the active loads and sized with approximately 75% of the aspect ratio of the active loads P1a and P1b, respectively.

Fig. 3.25 Electrical scheme of the proposed dynamic VC-biased OTA

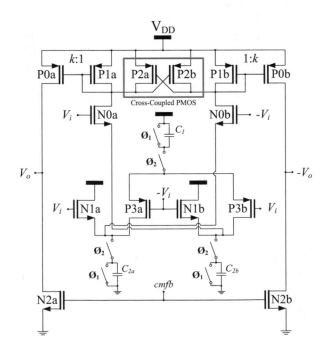

The dynamic biasing is controlled by three SC branches, comprising capacitors C_1, C_{2a}, and C_{2b}, which are charged during the amplification phase $\emptyset_2$ and partially discharged during the reset phase $\emptyset_1$. This way, the OTA starts slewing with a peak current, i.e., maximum GBW and peak slew-rate, and reaches the peak gm/I ratio, i.e., finite gain, in the end of phase $\emptyset_2$. The gain of a single-ended VC block in a continuous-time domain is in (3.34). The gain of the differential pair considering the cross-coupled load is shown in (3.35). Considering that gds_{N0} is sufficiently small when compared to gds_{P2}, a good approximation for the gain expression is given by (3.36). The body effect of the devices is neglected.

The gain of the amplifier is, as described in (3.34), directly proportional to the output conductance of the cross-coupled structure and inversely proportional to the square value of the latter. If a higher transconductance of P2 is delivered, the gain is also enhanced through the implementation of the auxiliary latch structure. A dynamically biased OTA is often designed firsthand, considering a continuous-time analysis domain, and afterward the implementation of the SC structures is carried out. Indeed, the sizing of the SC biasing branches relies on the static current consumption of active current sources, using (3.37), where R_{EQ} is the equivalent resistance of an active current source for a given sampling clock frequency. In the case of the work presented, a nominal frequency of 100 MHz is considered. The design of the OTA proposed in this section is first carried out in a continuous-time domain, targeting high gain and energy efficiency, by means of FOM as defined in (2.14). However, the operation of a dynamic OTA is inserted in a discrete-time context. Therefore, the expression in (2.14) must be modified into (3.38). The performance of the OTA proposed in this section is, consequently, evaluated in the terms of the FOM given by (3.38).

$$A_{v_{VC}} \approx 1 + \frac{gm_{P3}}{gm_{N1}} \Rightarrow |A_{v_{VC}}| > 1 \tag{3.34}$$

$$A_{v_{Latch}} = \frac{gm_{N0} \times gds_{P2}}{gds_{N0} \times gds_{P2} + gds_{P2}^{2} - gm_{P2}^{2}} \tag{3.35}$$

$$A_{v_{OTA}} \approx \frac{k \times (1 + A_{v_{VC}}) \times \frac{gm_{N0} \times gds_{P2}}{gds_{P2}^{2} - gm_{P2}^{2}}}{gds_{P0} + gds_{N2}} \tag{3.36}$$

$$R_{EQ} = \frac{1}{C_s \times f_s} \tag{3.37}$$

$$FOM = \frac{GBW \times C_L}{I_{DD}} \left[\frac{MHz \times pF}{mA_{RMS}} \right] \tag{3.38}$$

Following the sizing strategy in a continuous-time domain depicted in Fig. 3.26, targeting a gain higher than 50 dB, a GBW higher than 30 MHz, and a PM higher than 45° with a current consumption of less than 1 mA_{rms}, the circuit can be sized. The functional specifications considered in the sizing of the circuit are related to the

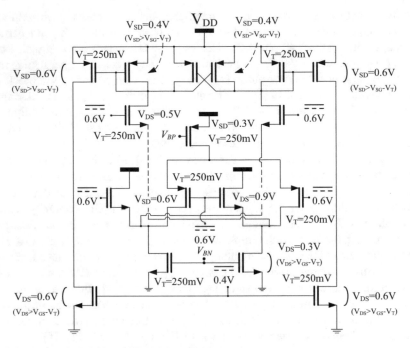

Fig. 3.26 Dynamic VC-biased OTA sizing strategy

overdrive voltages of the devices and to their minimal saturation margins, while the performance specifications are related with the performance metrics of the circuit. Moreover, the open-circuit test bench shown in Fig. 3.13 is used. The devices are set to operate in moderate inversion with an adequate saturation margin, i.e., more than 50 mV, to work as linearly as possible when needed, i.e., during phase $\emptyset_2$. Again, the GBW is evaluated considering a 6 pF load. Moreover, a minimum value of 3 mV for the systematic offset voltage is set, comprising 2.5% of the supply voltage, i.e., 1.2 V, and to avoid fabrication process and mismatch impact, all the active areas (AA) of the devices, i.e., the thin oxide areas and width-per-finger (WF) values, are taken into account. Finally, the maximum spectral density defined for acceptance is 30 nV/√Hz extracted at the value of GBW. In this case there is no body effect in any of the devices of the amplifier, since the fabrication of a prototype is out of the scope of this work, due to the difficult measurement as a stand-alone circuit. In fact, the complete test of the OTA proposed in this section must be embedded either in a SC filter or in an ADC for experimental evaluation purposes, due to the difficulties that arise when a real-time test bench is setup for evaluation of discrete-time measurements. Following (3.37) and taking into consideration that $V_{DSP2} = 245$ mV, $V_{DSN2} = 310$ mV, $I_{DSP2} - 2.5$ μA, and $I_{DSN2} = 105$ μA, it is possible to size the capacitors as $C_1 = 103.8$ fF and $C_2 = 3.4$ pF, for a 100 MHz sampling frequency. The AC response during $\emptyset_2$ is presented in Fig. 3.27, depicting a gain of 54.2 dB with a PM of 52.6°. During the reset phase, the quiescent current consumption is

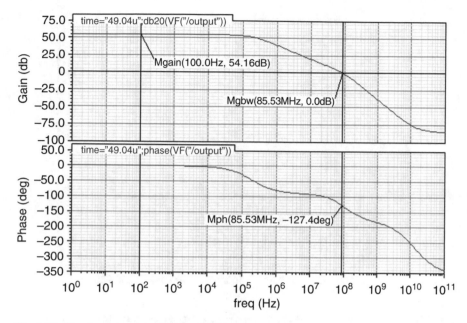

Fig. 3.27 Simulated dynamic VC-biased OTA AC response

only 348.3 µA, while in the amplification phase, the current consumption achieves the early peak value of 15.431 mA. The overall DC equivalent power, i.e., root mean square, consumption in each clock cycle, for a 1.2 V source, is 0.593 mW. The current consumption, i.e., I_{DD}, is shown in Fig. 3.28. A simulated FOM of 1038.8 MHz × pF/mA$_{rms}$ is achieved, with a 14.5 µV$_{rms}$ noise with a spectral density of 38.08 nV/√Hz extracted at 85.53 MHz, as shown in Fig. 3.29. The augmentation of noise, when compared to the static biasing, is mainly related to the smaller current employed to bias the differential pair, in the end of the amplification phase, i.e., phase $\emptyset_2$.

For proof of concept, a flipped-around sample-and-hold (S/H) simulation test bench for time-domain analyses, presented in Fig. 3.30, is implemented. The discrete Fourier transform (DFT) of the simulated response to a 4 MHz sinusoidal input with 100 mV$_{pp}$ amplitude is shown in Fig. 3.31. Coherent sampling is considered, using a Hamming truncation window and a 100 MHz clock frequency with 4096 samples, i.e., an accuracy of 12 bit, with no clock overlap. The DFT of the output signal shows a fundamental component approximately 75 dB better than the third harmonic which has a power of approximately −100 dB, as depicted in Fig. 3.31. A settling time of 72 ns is achieved for a 400 mV$_{pp}$ input step. Surveying the concept of coherent sampling, it is important to consider the fast Fourier transform (FFT), which is a common mathematical tool, useful to investigate performance of several sampled circuits and systems. Coherent sampling refers to a certain relationship between an input frequency, a sampling frequency, a number of cycles in the sampled set, and the number of samples itself. With coherent sampling, it is guaranteed that the signal

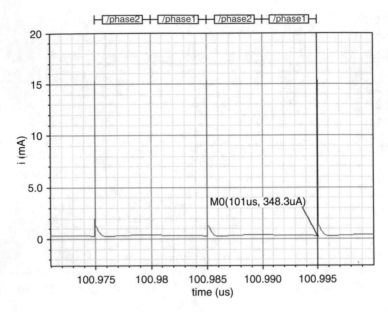

Fig. 3.28 Simulated dynamic VC-biased OTA current consumption during two clock cycles

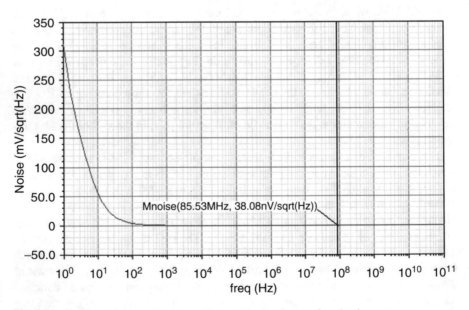

Fig. 3.29 Simulated dynamic VC-biased OTA equivalent input-referred noise response

Fig. 3.30 Flipped-around
S/H test bench for time-
domain analyses of the
dynamic VC-biased OTA

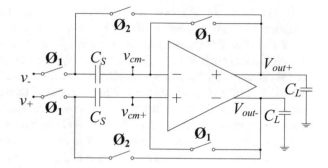

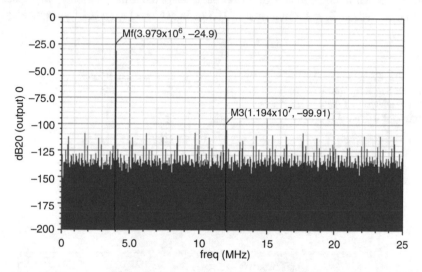

Fig. 3.31 Simulated dynamic VC-biased OTA DFT response to a 4 MHz and 100 mV$_{pp}$ sinusoidal
input

power in an FFT is contained within one FFT bin, considering a single input
frequency. The condition for coherent sampling is given by the equal quotients
between the input over the sampling frequencies and the number of cycles over the
number of samples. Yet, the number of samples value must be a power of 2, i.e., 2^{bits},
and the number of cycles must be an odd or prime number, e.g., 157, 163, 167, or
173. In this case, the number 163 is selected, resulting in an effective input frequency
of 3.9794 MHz, replacing the harsh value of 4 MHz.

A single-ended implementation, following the expression in (3.37) and taking into
consideration that $V_{DSP2} = 212$ mV, $V_{DSN2} = 320$ mV, $I_{DSP2} = 3.4$ µA, and
$I_{DSN2} = 43.4$ µA, is presented further. In this case, the capacitor values are
$C_1 = 160.5$ f. and $C_2 = 1.358$ pF, following (3.37). The AC response, in terms of
gain and phase during $\emptyset_2$, presents a gain of 50 dB with a less comfortable PM of

46.6° yet acceptable for most applications. During $\emptyset_1$ the quiescent current consumption is 123.8 μA, while in $\emptyset_2$ the current consumption achieves a peak of 15.58 mA. The DC equivalent power consumption, i.e., root mean square (RMS), in each clock cycle is 301 μW_{rms}. The noise response depicts a 35 μV_{rms} noise with a spectral density approximately 91.8 nV/√Hz extracted at GBW. A simulated FOM of 792 MHz × pF/mA$_{rms}$ is achieved. When compared to the fully differential implementation, the single-ended implementation enhances the noise response, due to the smaller current consumption. As before, an S/H test bench for time-domain analyses of the proposed single-ended implementation is shown in Fig. 3.32. Coherent sampling is considered also in this case, using a Hamming window and a 100 MHz clock frequency with 4096 samples, i.e., 12 bit accuracy. The DFT of the output in Fig. 3.33 shows a fundamental component 75 dB better than the third harmonic, which has a power of −98.6 dB. A settling time of 105 ns is achieved for a 0.4 V_{pp} input step.

Fig. 3.32 Flipped-around S/H test bench for time-domain analyses of the single-ended dynamic OTA

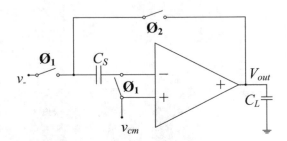

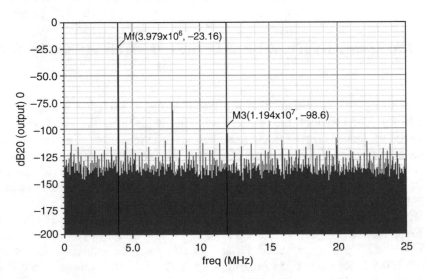

Fig. 3.33 Simulated single-ended dynamic OTA DFT response to a 4 MHz and 100 mV$_{pp}$ sinusoidal input

It is important to consider that the single-ended configuration maintains the even order harmonics, while the differential amplifier provides the inherent even order harmonics canceling. Indeed, a single-ended signal, which is unbalanced by definition, is evaluated by the difference between the signal of interest and a constant reference point, which is normally ground and serves as the return path for the signal. One handicap of this can be encountered if an error source is introduced into the signal path, because the ground reference will be unaffected by the injected error, and, therefore, the error is carried forward through the signal. Any signal variation introduced in a single-ended configuration will be difficult to remove without using overly complex cancelation techniques.

On the other hand, differential signals are made up of pairs of balanced signals moving at equal but opposite amplitudes around a reference point. The difference between the positive and negative balanced signals corresponds to the composite differential signal. If an error is introduced to a differential system path, the same will be added to each of the two balanced signals equally. Because the return path is not a constant reference point, the error will be canceled in the differential signal. Thus, differential signal chains are less susceptible to noise and interference. This inherent error cancelation also provides better CMRR and, in general, also better PSRR. The key performance parameters of both implementations are summarized in Table 3.4. When compared with the two-stage OTA in [10], it is possible to verify that the proposed solution has a higher FOM, with considerably less consumption than the fully differential implementation. The work presented in [10] shows a fairly low FOM of 128.7 MHz × pF/mA, with a current consumption of 23 mA. In this section, a fully dynamic voltage-combiner-biased OTA is addressed at analytical and behavioral level, with both single-ended and fully differential implementations. The proposed topology is, generically, a fully dynamic implementation of the folded voltage-combiner-biased OTA with gain enhancement through active PMOS loads in cross-coupled configuration. The implementation is carried out at sizing level and compounded at simulation level.

Table 3.4 Simulated key performance parameters of the dynamic VC-biased OTA

Metrics	Proposed OTA (Single-ended)	Proposed OTA (Fully differential)
Gain [dB]	50.1	54.2
Phase margin [°]	46.6	52.6
GBW at 6 pF [MHz]	33.13	85.53
VOS [mV]	0.98	3.05
RMS current [mA$_{rms}$]	0.251	0.494
Noise [μV$_{rms}$]	35	14.5
Noise density [nV/√Hz]	91.83	38.08
FOM [MHz × pF/mA$_{rms}$]	792	1038.8

3.6   Noise Modeling

In the next paragraphs, the attention turned toward the noise modeling of the elemental structures within the context of the proposed new family of CMOS amplifiers. Initial considerations and definitions are provided first; the most common types of noise that have impact in modern technologies are addressed second; and finally, the noise modeling is applied to the structures that compose this work.

3.6.1   Noise Definition

In electronics, noise is generically defined as an unwanted disturbance in an electrical signal. In other words, noise is every signal that does not contain or represent useful information for the user. Therefore, the study of this metric is important, since it can introduce levels and boundaries for the important signal, in order to avoid unwanted loss of information. The noise that is generated by electronic devices varies significantly, as it is produced by different sources and results in diverse effects. Moreover, in communication systems, noise is an error or undesired random disturbance of a useful information signal. The noise is, therefore, a summation of unwanted or disturbing energy from natural and sometimes man-made sources. In a given electronic circuit, even if all the external sources of noise are eliminated, there is always a level of noise at the output of the circuit. This is due to the internal sources of noise, which are inherent to electronic components and devices. Moreover, the analysis and summation of multiple noise sources are usually carried in terms of power spectral density, in the frequency domain, and often measured in V^2/Hz.

3.6.2   Noise Types

Different types of noise are generated by different devices and different processes. Thermal noise is unavoidable at non-zero temperature, while other types depend mostly on device type, as in the case of the shot noise, which needs a steep potential barrier and is mainly present in bipolar technologies, or manufacturing quality and semiconductor defects, such as conductance fluctuations, e.g., flicker noise. The most common noise types in literature are briefly overviewed in the next paragraphs: the flicker noise, the thermal noise, the shot noise, and, finally, the burst noise.

The flicker noise, also known as $1/f$ noise, is a signal or process with a frequency spectrum that falls off steadily into the higher frequencies, with a pink spectrum. This type of noise occurs in almost all electronic devices and results from a variety of effects, mainly related with traps and defects in crystal lattice, e.g., in semiconductors or carbon structures.

The thermal noise is generated by the random thermal motion, i.e., random fluctuations of velocity of charge carriers, e.g., electrons, fighting for space inside an electrical conductor, which happens regardless of any applied voltage. This type of noise and the corresponding effects are predominant in MOSFET technologies and also in resistors. The thermal noise has, generically, a white spectrum, in the sense that its power spectral density is nearly equal throughout the frequency spectrum. The amplitude of the noise signal has very nearly a Gaussian probability density function [11].

The shot noise results from random statistical fluctuations of the electric current when the charge carriers, e.g., electrons, traverse a gap. In other words, it is due to a DC current that flows through a p-n junction. If electrons flow across a barrier, then they have discrete arrival times. Those discrete arrivals display shot noise. Typically, the barrier in a diode is used. Shot noise is similar to the noise created by rain falling on a tin roof. The flow of rain may be relatively constant, but the individual raindrops arrive discretely. The RMS value of the shot noise current I_{sh} is given by the Schottky formula set by (3.39), where I is the DC current, q is the charge of an electron, and Δf is the integrating bandwidth [12]. The Schottky formula assumes independent arrivals and is valid until the frequency is comparable with $1/t$, where t represents the transit time in the depletion region of the junction. The amplitude of the noise signal has very nearly a Gaussian probability density function. The noise spectral density can be given by (3.40), which relates with (3.39) through the removal of the square root and the bandwidth integrating factor, i.e., Δf.

The burst noise is originated by heavy metal ion contamination and consists of sudden transitions between two or more levels, i.e., non-Gaussian, in the order of several hundred microvolts, at random and unpredictable times. Each shift in offset voltage lasts for several milliseconds, and the intervals between pulses tend to be in the audio range, i.e., less than 100 Hz, hence the term popcorn noise for the crackling sounds it produces in most audio circuits.

$$I_{sh} = \sqrt{2qI\Delta f} \tag{3.39}$$

$$\overline{I_n^2} = 2qI \tag{3.40}$$

3.6.3 Generic Modeling

The flicker noise can be observed in almost every electronic device, from homogeneous resistors to semiconductor devices and even in chemical concentrated cells. Due to the fact that flicker noise is well spread over the components, engineers often think that there is a key physical mechanism behind it. However, until now, such a mechanism has not yet been found. In truth, all theories and models differ in detail, yet all are based on the mobility fluctuation model expressed by the Hooge empirical relation, and the carrier density or number fluctuation model, first introduced by

Fig. 3.34 Thermal noise
model in a resistor: (**a**)
voltage source in series; (**b**)
current source in parallel

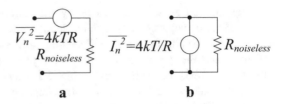

McWhorter [13, 14]. Moreover, the average power of the flicker noise cannot be predicted easily: it varies, depending on the cleanness of oxide-silicon interface from one CMOS technology to another. Despite all this issues, it is well established in most literature that the flicker noise is more easily modeled as a voltage source in series with the gate; if the MOSFET is in the saturation region, the noise spectral density is roughly given by (3.40), where K is a process-dependent constant, in the order of 10^{-25} V^{2}F, C_{ox} is the oxide capacitance in MOSFET devices, and W and L are channel width and length, respectively [11]. This is an empirical model, commonly used in literature [15].

The thermal noise has its origin in the thermal agitation of carriers inside a conductor and imposes a minimum level for the noise inside an electronic circuit. In all conductors, there is a nonperiodic voltage with amplitude that is proportional to the temperature. Starting with a simple practical example, the noise spectral density that is present in a resistor R is given by (3.41), where k is the Boltzmann constant, and T is the absolute temperature in °K. It is possible to define the thermal noise in a resistor, following the circuit in Fig. 3.34. In fact, it is possible to define the noise contribution of a resistor, by placing a voltage source in series or a current source in parallel, with the noiseless model of the first. The value of the noise that is present in a resistor R can also be given by (3.42). The proper choice of the noise source can ease the circuit analysis. In order to reduce the impact of the thermal noise in an electronic circuit, it is useful to reduce the amount of resistors that are employed. Moreover, the noise modeling has no actual direction, since it is due to a motion that is random in nature. The RMS value of noise can be obtained by applying a square root to the integration of the noise spectral density over a bandwidth, as given by (3.43) [11].

$$\overline{V_n^2} = \frac{K}{C_{ox}WL} \times \frac{1}{f} \tag{3.40}$$

$$\overline{V_n^2} = 4kTR \tag{3.41}$$

$$\overline{I_n^2} = \frac{4kT}{R} \tag{3.42}$$

$$V_{n_{RMS}} = \sqrt{\int_{-\infty}^{+\infty} (4kTR)df} = 4kTR\Delta f \qquad (3.43)$$

In a MOSFET, the most important noise type, i.e., the noise that has the most important impact, is the thermal noise. A MOSFET has, in normal working order, an inverse resistive channel between the drain and the source. The gate voltage forms with minority carriers the channel. In the extreme case when the drain-source voltage $V_{DS} = 0$ V, the channel can be treated as a homogeneous resistor. The noise spectral density in the channel is given (3.44), which derives directly from (3.42), in which the resistor value is replaced by r_o.

$$\overline{I_n^2} = \frac{4kT}{r_o} = 4kT\text{gds} \qquad (3.44)$$

However, normally the voltage $V_{DS} \neq 0$ V, and through experimentation (3.45) is a fairly good approximation, as long as the device is well saturated. The reason for this is that the cutoff region nearby the drain is much smaller than the resistive reverse channel which is responsible for the noise. The expression in (3.45) predicts the thermal noise in the channel without the substrate effect. In practice the thermal noise is higher. The factor γ is a complex function of the basic transistor parameters and bias conditions. For modern CMOS processes with oxide thickness t_{ox} in the order of 50 nm and with a lower substrate doping between 10^{15} and 10^{16} cm^{-3}, the factor γ is between 0.67 and 1; hence, the value 2/3 is considered throughout the next paragraphs. Considering the gate resistance, it is possible to define the noise contribution of this terminal with a noise voltage source that follows the noise contribution of a simple resistor, as in (3.41). However, this value can be neglected if $r_g/3 \gg \gamma/\text{gm}$. This simplification is taken into account further. The thermal noise sources in a MOSFET can be modeled following the circuit in Fig. 3.35.

$$\overline{I_n^2} = 4kT\gamma\text{gm} = 4kT\frac{2}{3}\text{gm} \qquad (3.45)$$

Considering the simple CS configuration in Fig. 3.36, the output noise can be derived following (3.46). It is important to keep in mind that the output resistance of a MOSFET is not an actual resistance, i.e., it is simply a small-signal equivalent device; thus, the noise contribution does not follow (3.42). At this point, the input-

Fig. 3.35 Thermal noise model in a MOSFET

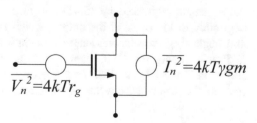

$\overline{V_n^2} = 4kTr_g$

$\overline{I_n^2} = 4kT\gamma\text{gm}$

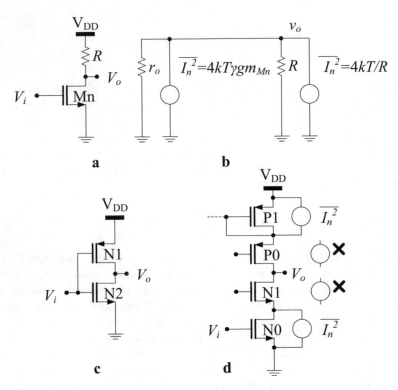

Fig. 3.36 Noise model of basic topologies: (**a**) CS; (**b**) CS small-signal; (**c**) CMOS inverter; (**d**) cascode stage

referred noise can be defined. For this purpose, it is enough to divide the output noise by the squared gain of the topology as in (3.47). In practice, the input-referred noise is the equivalent noise value that should be applied at the input of a corresponding noiseless topology, to obtain the exact same value of noise at the output of the noisy one. In closed-circuit configurations, the contributions of R and r_o can be neglected, turning (3.47) into (3.48), where the parameter Γ is the actual noise contribution of the MOSFET and equals 1 in this case. This parameter is useful to compare the noise degradation of more complex circuits, when compared to a single device. Considering the CMOS inverter in Fig. 3.36, and following the same type of analysis, the input-referred noise can be defined by (3.49), and if $gm_{N1} \approx gm_{N2} \approx gm$, (3.49) turns into (3.50), and Γ equals 2. In the cascode stage of Fig. 3.36, the input-referred noise can be expressed by (3.51) and (3.52). It is assumed that N2 contributes negligibly to the noise at the output, which is valid especially at low frequencies and neglecting the channel-length modulation of N1. In this last case, Γ equals 2. Finally, using the CMOS inverter in an inverting amplifier configuration, as shown in Fig. 3.37, the RMS output noise can be extracted by applying the square root to the product between three terms: the input-referred noise, the desired

Fig. 3.37 CMOS inverter
in inverting amplifier
configuration

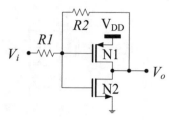

bandwidth, and the squared configuration gain, as expressed in (3.53), following the definition in (3.47).

$$\overline{V_{\text{nout}}}^2 = \left(\gamma gm_{\text{Mn}} + \frac{1}{R}\right)(4kT)(r_o\|R)^2 \tag{3.46}$$

$$\overline{V_{\text{nin}}}^2 = \frac{\overline{V_{\text{nout}}}^2}{A_v{}^2} = \frac{\overline{V_{\text{nout}}}^2}{(gm_{\text{Mn}}(r_o\|R))^2} \tag{3.47}$$

$$\overline{V_{\text{nin}}}^2 = \frac{4kT\gamma}{gm_{\text{Mn}}} \times 1 = \frac{4kT\gamma}{gm_{\text{Mn}}} \times \Gamma; \Gamma = 1 \tag{3.48}$$

$$\overline{V_{\text{nin}}}^2 = \frac{4kT\gamma}{gm_{\text{N1}}} + \frac{4kT\gamma}{gm_{\text{N2}}} \tag{3.49}$$

$$\overline{V_{\text{nin}}}^2 = \frac{4kT\gamma}{gm} \times 2 = \frac{4kT\gamma}{gm} \times \Gamma; \Gamma = 2 \tag{3.50}$$

$$\overline{V_{\text{nin}}}^2 = \frac{4kT\gamma}{gm_{\text{N0}}} + \frac{4kT\gamma gm_{\text{P1}}}{gm_{\text{N0}}{}^2} = \frac{4kT\gamma}{gm_{\text{N0}}} \times \left(1 + \frac{gm_{\text{P1}}}{gm_{\text{N0}}}\right) \tag{3.51}$$

$$\overline{V_{\text{nin}}}^2 = \frac{4kT\gamma}{gm} \times \left(1 + \frac{gm}{gm}\right) = \frac{4kT\gamma}{gm} \times 2 = \frac{4kT\gamma}{gm} \times \Gamma; \Gamma = 2 \tag{3.52}$$

$$V_{\text{noutRMS}} = \sqrt{\overline{V_{n_{\text{in}}}}^2 \times \Delta f \times A_v{}^2} = \sqrt{\overline{V_{n_{\text{in}}}}^2 \times \Delta f \times \left(\frac{R2}{R1}\right)^2} \tag{3.53}$$

3.6.4 Voltage-Combiner Noise Modeling

Following the same circuit analysis, the important noise contributions in a VC structure are shown in Fig. 3.38. The most important noise sources are modeled as drain-source current sources. The output noise can be calculated as in (3.54). Using

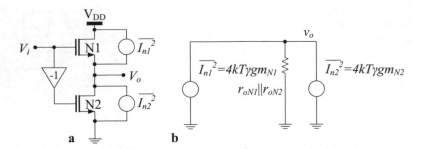

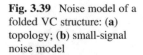

Fig. 3.38 Noise model of a VC structure: (**a**) topology; (**b**) small-signal noise model

Fig. 3.39 Noise model of a folded VC structure: (**a**) topology; (**b**) small-signal noise model

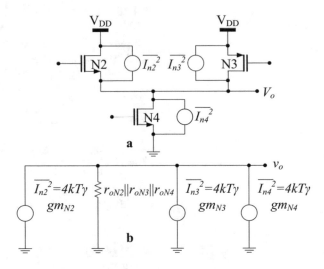

the noise contributions given by (3.55) and (3.56), the input-referred noise is given by (3.57), and considering that $gm_1 \approx gm_2 \approx gm$, it is fair to approximate the noise impact, Γ, of the VC structures, to a value close to 2. Note that the contributions of the small-signal output resistances are neglected in the input-referred noise expressions. Moreover, the body effect of N1 is neglected for simplicity. Regarding the folded VC structure, and following the same type of circuit analysis, it is possible to acknowledge the most important noise contributions, which are detailed in Fig. 3.39. The output noise can be calculated following (3.58). Using the noise contributions given by (3.59), (3.60), and (3.61), the input-referred noise is given by (3.62), and considering that $gm_2 \approx gm_3 \approx gm_4 \approx gm$, it is fair to approximate the noise impact, Γ, of the VC structures, to a value close to 3. Note that the contributions of the small-signal output resistances are neglected in the input-referred noise expressions. Moreover, the body effect of N2 is neglected for simplicity. The output noise of the VC-biased OTA can be extracted using the previous detailed approach and considering the noise contribution of a differential pair device, as illustrated in Fig. 3.40. The output noise current can be given by (3.63), translating into voltage

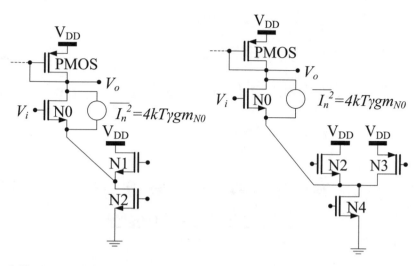

Fig. 3.40 Noise model with the differential-pair and PMOS current-mirror devices

through (3.64), where B is the current-mirroring factor, from the differential pair to the output branches. The same principle can be applied to determine the output noise of the folded VC biased without loss of detail. The output noise voltage of the folded VC-biased OTA uses (3.65) and is given by (3.66). With minor approximations, the input-referred noise of the OTAs can be given, respectively, by (3.67) and (3.68). If the top PMOS current-mirror noise contributions are considered, (3.67) becomes (3.69) and (3.68) becomes (3.70). It is possible to design the proposed OTAs for minimum Γ values, when noise is a relevant performance issue.

$$\overline{V_{nout}}^2 = \left(\overline{i_{n1}}^2 + \overline{i_{n2}}^2\right)\left(r_{o_{N1}} \| r_{o_{N2}}\right)^2 = (gm_{N1} + gm_{N2})(4kT\gamma)\left(r_{o_{N1}} \| r_{o_{N2}}\right)^2 \quad (3.54)$$

$$\overline{I_{nN1}}^2 = 4kT\gamma gm_{N1} \quad (3.55)$$

$$\overline{I_{nN2}}^2 = 4kT\gamma gm_{N2} \quad (3.56)$$

$$\overline{V_{nin_{N2}}}^2 = \frac{4kT\gamma gm_{N1} + 4kT\gamma gm_{N2}}{gm_{N2}^2} = \frac{4kT\gamma}{gm_{N2}} \times \Gamma; \Gamma = \left(1 + \frac{gm_{N1}}{gm_{N2}}\right) \quad (3.57)$$

$$\overline{V_{nout}}^2 = \left(\overline{i_{n2}}^2 + \overline{i_{n3}}^2 + \overline{i_{n4}}^2\right)\left(r_{o_{N2}} \| r_{o_{N3}} \| r_{o_{N4}}\right)^2$$
$$= (gm_{N2} + gm_{N3} + gm_{N4})(4kT\gamma)\left(r_{o_{N2}} \| r_{o_{N3}} \| r_{o_{N4}}\right)^2 \quad (3.58)$$

$$\overline{I_{nN2}}^2 = 4kT\gamma gm_{N2} \quad (3.59)$$

$$\overline{I_{nN3}}^2 = 4kT\gamma gm_{N3} \tag{3.60}$$

$$\overline{I_{nN4}}^2 = 4kT\gamma gm_{N4} \tag{3.61}$$

$$\overline{V_{nin_{N3}}}^2 = \frac{4kT\gamma gm_{N2} + 4kT\gamma gm_{N3} + 4kT\gamma gm_{N4}}{gm_{N3}^2} = \frac{4kT\gamma}{gm_{N3}} \times \Gamma; \Gamma$$

$$= \left(1 + \frac{gm_{N2}}{gm_{N3}} + \frac{gm_{N4}}{gm_{N3}}\right) \tag{3.62}$$

$$\overline{I_{nout}}^2 = B \times (gm_{N0} + gm_{N1} + gm_{N2})(4kT\gamma) \tag{3.63}$$

$$\overline{V_{nout}}^2 = \frac{B \times (gm_{N0} + gm_{N1} + gm_{N2})(4kT\gamma)}{gds_{out}^2} \tag{3.64}$$

$$\overline{I_{nout}}^2 = B \times (gm_{N0} + gm_{N2} + gm_{N3} + gm_{N4})(4kT\gamma) \tag{3.65}$$

$$\overline{V_{nout}}^2 = \frac{B \times (gm_{N0} + gm_{N2} + gm_{N3} + gm_{N4})(4kT\gamma)}{gds_{out}^2} \tag{3.66}$$

$$\overline{V_{nin}}^2 = \frac{(4kT\gamma)}{gm_{N0}} \times \left(1 + \frac{gm_{N1}}{gm_{N0}} + \frac{gm_{N2}}{gm_{N0}}\right) = \frac{(4kT\gamma)}{gm_{N0}} \times \Gamma, \Gamma$$

$$= \left(1 + \frac{gm_{N1}}{gm_{N0}} + \frac{gm_{N2}}{gm_{N0}}\right) \tag{3.67}$$

$$\overline{V_{nin}}^2 = \frac{(4kT\gamma)}{gm_{N0}} \times \left(1 + \frac{gm_{N2}}{gm_{N0}} + \frac{gm_{N3}}{gm_{N0}} + \frac{gm_{N4}}{gm_{N0}}\right) = \frac{(4kT\gamma)}{gm_{N0}} \times \Gamma, \Gamma$$

$$= \left(1 + \frac{gm_{N2}}{gm_{N0}} + \frac{gm_{N3}}{gm_{N0}} + \frac{gm_{N4}}{gm_{N0}}\right) \tag{3.68}$$

$$\overline{V_{nin}}^2 = \frac{(4kT\gamma)}{gm_{N0}} \times \left(1 + \frac{gm_{N1}}{gm_{N0}} + \frac{gm_{N2}}{gm_{N0}} + \frac{gm_{PMOS}}{gm_{N0}}\right) = \frac{(4kT\gamma)}{gm_{N0}} \times \Gamma, \Gamma$$

$$= \left(1 + \frac{gm_{N1}}{gm_{N0}} + \frac{gm_{N2}}{gm_{N0}} + \frac{gm_{PMOS}}{gm_{N0}}\right) \tag{3.69}$$

$$\overline{V_{nin}}^2 = \frac{(4kT\gamma)}{gm_{N0}} \times \left(1 + \frac{gm_{N2}}{gm_{N0}} + \frac{gm_{N3}}{gm_{N0}} + \frac{gm_{N4}}{gm_{N0}} + \frac{gm_{PMOS}}{gm_{N0}}\right)$$

$$= \frac{(4kT\gamma)}{gm_{N0}} \times \Gamma, \Gamma = \left(1 + \frac{gm_{N2}}{gm_{N0}} + \frac{gm_{N3}}{gm_{N0}} + \frac{gm_{N4}}{gm_{N0}} + \frac{gm_{PMOS}}{gm_{N0}}\right) \tag{3.70}$$

In this section, the generic definitions of the most common types of noise that can be found in modern technologies are provided, followed by typical noise models in MOSFET devices; and finally, the noise models for the VC structure, for the folded VC structure, for the VC-biased OTA, and for the folded VC-biased OTA are provided.

3.7 Summary

This chapter presents the topological and analytical descriptions of a new family of single-stage OTAs, biased by voltage-combiner structures, with high gain and energy efficiency. In this new family, the traditional current source that typically bias the differential pair of a symmetrical CMOS OTA is replaced by several configurations of voltage-combiners, i.e., in a cross-coupled configuration, in a folded configuration to tackle specifically low-voltage supplies, and finally in a dynamic configuration to be implemented in high-performance ADCs and SC filters. A summary of the works proposed and respective implementations is presented in Table 3.5. The folded VC-biased OTA and the dynamic VC-biased OTA present the highest values of energy efficiency, i.e., FOM, among the proposed topologies, which are already comparable with the state of the art. The high values of FOM are mainly related with the low-power consumption of these architectures. It is possible to acknowledge a clear evolution line, which shows an enhancement of the energy efficiency among the proposed topologies. In the next chapter, all the topologies proposed in this PhD work are optimized at device/sizing level, resulting in considerable improvements, both in terms of FOM and gain, which relocate the proposed family of amplifiers in the state of the art of single-stage amplifiers, thus demonstrating the potential of both the circuits and the optimization framework.

Table 3.5 Comparison among the proposed single-stage amplifiers

References	Gain [dB]	I_{DD} [mA]	PM [°]	GBW [MHz]	FOM [MHz × pF/mA]	Load [pF]	Tech [nm]
VC-biased OTA[a]	44.8	1.3	84.1	51.2	236.3	6	130
VC-biased OTA w/ current starving[a]	63.1	2.0	60.1	146.7	440.1	6	130
Folded VC-biased OTA[a]	40.2	0.4	73.3	67.55	1013.25	6	130
Dynamic VC-biased OTA SE[a]	50.1	0.251	46.6	33.13	792	6	130
Dynamic VC-biased OTA FD[a]	54.2	0.494	52.6	85.53	1038.8	6	130

[a]Simulation results before optimization

References

1. F. Leyn, et al., "A behavioral signal path modeling methodology for qualitative insight in and efficient sizing of CMOS OpAmps," in IEEE/ACM International Conference on Computer-Aided Design, Digest of Technical Papers, Page(s): 374–381, Nov. 1997. DOI: https://doi.org/10.1109/ICCAD.1997.643563.
2. F. Grasso, et al., "SapWin 4.0-a new simulation program for electrical engineering education using symbolic analysis," in Computer Applications in Engineering Education, Vol. 24, Issue 1, Page(s): 44–57, Jan. 2016. DOI: https://doi.org/10.1002/cae.21671.
3. R. Spence, R. Soin, "Tolerance Design of Electronic Circuits," Addison-Wesley, 1988. DOI: 9780201182422.
4. P. Ko, "Approaches to Scaling," in VLSI Electronics: Microstructure Science, Vol. 18, Issue 1, Page(s): 1–37, Academic Press, 1989. DOI: https://doi.org/10.1016/B978-0-12-234118-2.50005-X.
5. Online, 2017: https://www.ti.com/lit/ds/symlink/ths4524.pdf.
6. W. Sansen, "Analog Design Essentials," Springer, 2006. ISBN: 978-0-387-25747-1.
7. M. Copeland, J. Rabaey, "Dynamic Amplifiers for MOS Technology," in Electronics Letters, Vol. 15, Page(s): 301–302, May 1979. DOI: https://doi.org/10.1049/el:19790214.
8. B. Hosticka, "Dynamic CMOS Amplifiers," in IEEE Journal of Solid-State Circuits, Vol. 15, Issue 5, Page(s): 887–894, Oct. 1980. DOI: https://doi.org/10.1109/JSSC.1980.1051488.
9. B. Hosticka, et al., "Performance of Integrated Dynamic MOS Amplifiers," in Electronics Letters, Vol. 17, Issue 8, Page(s): 298–300, Apr. 1981. DOI: https://doi.org/10.1049/el:19810209.
10. A. Zadeh, "A 100MHz, 1.2V, ±1V Peak-to-Peak Output, Double-Bus Single Ended-to-Differential Switched Capacitor Amplifier for Multi-Column CMOS Image Sensors," in IEEE International New Circuits and Systems Conference (NEWCAS), Page(s): 1–4, Jun. 2016. DOI: https://doi.org/10.1109/NEWCAS.2016.7604739.
11. Z. Chong, et al., "Low-Noise Wide-Band Amplifiers in Bipolar and CMOS Technologies," Springer, 1991. ISBN: 978-1-4757-2126-3.
12. W. Schottky, "Über Spontane Stromschwankungen in Verschiedenen Elektrizitätsleitern," in Anallen der Physik, Vol. 362, Issue 23, Page(s): 541–567, 1918. DOI: https://doi.org/10.1002/andp.19183622304.
13. F. Hooge, "1/f Noise Sources," in IEEE Transactions on Electron Devices, Vol. 41, Issue 11, Page(s): 1926–1935, November 1994. DOI: https://doi.org/10.1109/16.333808.
14. L. Vandamme, et al., "On the Additivity of Generation-Recombination Spectra Part 3: The McWhorter Model for 1/f Noise in MOSFETs," in Physica B: Condensed Matter, Vol. 357, Issues 3–4, Page(s): 507–524, Elsevier, Mar. 2005. DOI: https://doi.org/10.1016/j.physb.2004.09.106.
15. C. Motchenbacher, et al., "Low-noise Electronic System Design," Wiley Interscience, 1993. ISBN: 0-471-57742-1.

Chapter 4
Design Optimization and Results

4.1 Optimization Framework

In order to achieve the most competitive design solutions and completely explore the trade-offs between the energy efficiency and the low-frequency gain in the design of the OTAs proposed in this work, a multi-objective and multi-constraint circuit optimization framework denominated AIDA (analog IC design automation) is utilized to automatically optimize the circuits at sizing level [1].

The optimization framework AIDA is based on an evolutionary computation kernel and implements an automatic synthesis flow, where the robustness of the solutions is enhanced by considering process, voltage, and temperature (PVT) corners and the circuit performance is evaluated using electrical circuit simulators. In this work, the Eldo® simulator is used. It is important to consider that minor discrepancies between simulators are always expected, since the technology models are themselves also marginally dissimilar. The robustness of the design is also enhanced by considering in-circuit Monte Carlo yield optimization [2]. The output of AIDA is a Pareto optimal front (POF) of feasible circuit solutions, i.e., a set of sized circuits all meeting the specifications but representing different compromises between the optimization objectives, providing a large insight of the circuit.

The complete AIDA framework, depicted in Fig. 4.1, is composed of two main modules: a sizing optimization module, i.e., AIDA-C [3], and a layout generation module, i.e., AIDA-L [4], which are not considered in this work. The circuit sizing module is detailed in the next paragraph. The layout generation module can be partially activated to include layout data, e.g., geometric and parasitic, in the sizing optimization flow, for layout-aware sizing optimization. The layout generator, i.e., AIDA-L, takes the device sizes and the best floor plan for selected circuit sizing and generates the complete layout by placing and routing the devices. After the circuit-level design, the layout generator creates the layout by placing and routing the devices. The generated circuit layout is then saved as a geometric data stream

© Springer International Publishing AG, part of Springer Nature 2019
R. F. S. Póvoa et al., *A New Family of CMOS Cascode-Free Amplifiers with High Energy-Efficiency and Improved Gain*,
https://doi.org/10.1007/978-3-319-95207-9_4

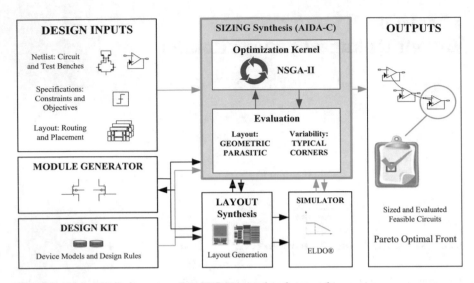

Fig. 4.1 Analog IC design automation (AIDA) complete framework

(GDS-II) graphical format, and the physical evaluation of the results is performed. In addition, an analog module generator (AMG) is used by both the modules, i.e., sizing and layout modules. This AMG extends the degrees of freedom explored during the synthesis, providing a wide range of layout options for each device. Ultimately, it is provided to the designer running the framework the freedom to choose which solution fits best the requirements of a given design project and, evidentially, targeting a specific application.

The circuit-level sizing optimization framework, AIDA-C, receives the circuit and test benches in netlist format with free variables, i.e., design parameters, to be optimized automatically; a complete set of performance and functional specifications, e.g., gain and overdrive voltages correspondingly; and also the user-defined optimization objectives, i.e., specifications to be maximized or minimized, e.g., maximization of GBW and minimization of current consumption, that often represent a design trade-off. The AIDA-C optimization engine solves multi-objective multi-constraint optimization problems as defined in (4.1), where x is a vector of N input variables to optimize, $f_m(x)$ is a set of M objective functions to minimize, $g_j(x)$ is the set of J constraints to be accomplished, and finally, $x_i^L \leq x_i \leq x_i^U$ is the range of the optimization variable x_i.

$$
\begin{aligned}
\text{find } x \text{ that minimizes } f_m(x) \quad & m = 1, 2, \ldots M \\
\text{subject to } g_j(x) \geq 0 \quad & j = 1, 2, \ldots J \\
x_i^L \leq x_i \leq x_i^U \quad & i = 1, 2, \ldots N
\end{aligned} \tag{4.1}
$$

This general formulation is adapted to the analog IC design problem by using the circuit parameters as ranged design variables, defining the N dimensions real search space. The objectives are handled as defined in (4.2), where the evaluated performance figures, p_m, to be minimized are used directly as one of the $f_m(x)$ objective,

Fig. 4.2 Simplified pseudo-code of the NSGA-II algorithm in AIDA-C

input:		Population Size, Max Generation
1.		initialization
2.		**while** (generation < Max Generation) {
3.		apply operators //Mutation and Crossover
4.		evaluate
5.		non-dominated sorting
6.		selection
7.		}
8.		**return** pareto

while the ones being maximized are multiplied by -1, for simple normalization. The design constraints are normalized according to (4.3), where p_j are the evaluated circuit characteristics and P_j is the constraint limit, as defined in detail in [5]. The multi-objective optimization kernel implements the NSGA-II algorithm. The implementation of the NSGA-II algorithm follows the reference first proposed in [6], using simulated binary crossover and mutation operators, tournament selection, and constrained based solution dominance check, as briefly overviewed in Fig. 4.2.

$$f_m(x) = \begin{cases} p_m & \text{when minimizing } p_m \\ -p_m & \text{when maximazing } p_m \end{cases} \tag{4.2}$$

$$g_i(x) = \begin{cases} \dfrac{p_i - P_i}{|P_i|} & \text{when the constraint is } p_i \geq P_i \\ p_i & \text{when the constraint is } p_i \geq 0 \\ -p_i & \text{when the constraint is } p_i \leq 0 \\ \dfrac{P_i - p_i}{|P_i|} & \text{when the constraint is } p_i \leq P_i \end{cases} \tag{4.3}$$

4.2   Voltage-Combiner Biased OTA

The optimization of the amplifiers, in this chapter, is carried out using UMC 130 nm CMOS design kit, with standard-VT 3.3 V devices. For the implementation in a fully differential configuration, a continuous-time common-mode feedback circuit is incorporated in the circuit, similar to the one presented in Fig. 4.3. Theoretically, this circuit compares the output DC voltage with half of V_{DD}, setting the output on approximately 1.65 V through signal reinjection in the cmfb node. Moreover, a capacitor can be employed between n_{min} and ground, to ease supply noise.

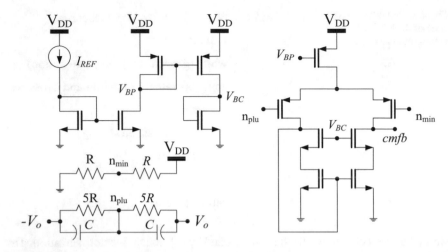

Fig. 4.3 Electrical scheme of the common-mode feedback circuit

In order to achieve the most competitive design solutions and completely explore the trade-offs between the energy efficiency, in terms of current consumption versus settling speed, and the low-frequency gain of the proposed OTA, a multi-objective multi-constraint automatic IC sizing and optimization framework, AIDA-C [1], is used to automatically synthesize the circuit at sizing level. Note that, at this point, the common-mode feedback circuit shown in Fig. 4.3 is employed, targeting the fabrication of an optimized integrated prototype.

Simulation results of a properly optimized circuit, using AIDA-C [1], demonstrate that low-frequency gains above 47 dB can be achieved over corner conditions, as described further. The optimization process is carried out considering two contradictory objectives: the maximization of the low-frequency gain and the maximization of the FOM, as given in (2.14), where C_L is the load capacitance and I_{DD} is the current consumption, not considering external biasing sources, in this case I_{REF}. This resulted in not one but several sizing solutions, constituting a POF of valid sized circuits, as described further. The extension to deeper nanoscale nodes, e.g., 65 nm or 40 nm, and lower supply voltages, e.g., 2 V, is relatively straightforward. For supply voltages of 1.2 V or below, the voltage combiners need to be implemented in a folded configuration, described ahead, in order to properly bias the CMOS devices in DC. In addition to the objectives, a set of performance and functional specifications are also considered in the optimization process. The functional specifications are related to the overdrive voltage (OVD) of the devices and to their minimal saturation margin (DEL), i.e., V_{DS} minus VDS_{SAT}, while the performance specifications are related with the performance metrics of the circuit, e.g., low-frequency gain, phase margin, or GBW. The devices are set to operate in moderate inversion, with a comfortable saturation margin, to work as linearly as possible when needed and avoid distortion. Nevertheless, since the implementation is in a single-well configuration, the critical devices N0 and N1 pairs suffer from body effect. All the specifications are shown in Table 4.1. The GBW calculation took a 6 pF load into

Table 4.1 VC biased OTA optimization performance and functional constraints

Metrics	Specifications	Metrics	Specifications
GBW	$\geq$30 MHz	OVD	$\geq$0.05 V
Gain	$\geq$45 dB	DEL	$\geq$0.07 V
I_{DD}	$\leq$2 mA	AA	$\geq$1 μm^2
PM	$\geq$55°	WF	$\geq$0.72 μm
FOM	$\geq$900 MHz × pF/mA	VOS	$\leq$5 mV

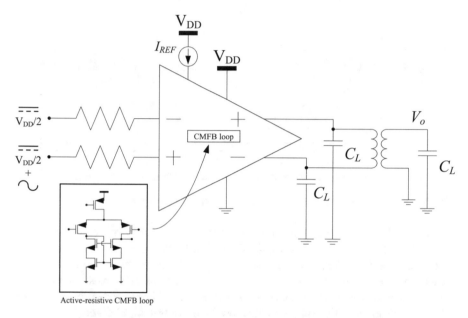

Active-resistive CMFB loop

Fig. 4.4 Open-circuit test bench circuit with common-mode feedback circuit

consideration. Moreover, a minimum value of 5 mV for the offset voltage is set, i.e., 0.15% of the biasing voltage, and in order to avoid fabrication mismatch impact, all active areas and width-*per*-finger values are accounted. In fact, the AAs and WFs are set to have more than 1 μm^2 and 0.72 μm, correspondingly, to incorporate the aspect ratio in the design process, enhancing the robustness to process mismatch.

The optimization variables are the channel widths, channel lengths, and the number of fingers of the devices. A CMFB circuit, similar to the one presented in Fig. 4.3 of the present chapter, is also considered in the optimization process, with functional specifications according. This circuit compares the output DC voltage with half of V_{DD}, in this case 1.65 V, setting the output voltage on approximately 1.65 V with signal reinjection into the cmfb node. The optimization is carried out considering worst case corner conditions: slow-n/slow-p, slow-n/fast-p, fast-n/slow-p, fast-n/fast-p device models of the UMC 130 nm design kit. A simulation temperature of 27 °C, a fixed nominal supply voltage at 3.3 V, and a 1.65 V common-mode voltage are considered in this optimization process. The optimization of the circuit is carried out using an open-circuit test bench, similar to the one presented in Fig. 4.4, allowing the

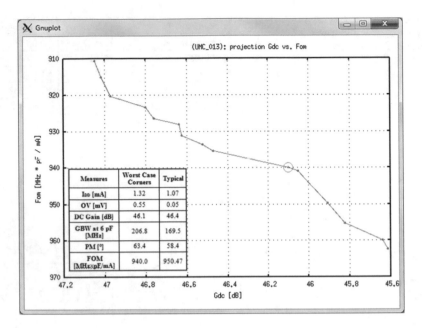

Fig. 4.5 VC biased OTA optimization POF considering corner conditions

extraction of the low-frequency gain, the GBW, phase margin, and also the systematic offset voltage. The presented optimization takes approximately 4 h in an i7-3770 Intel© CPU with 16 GB RAM. A population of 128 elements evolving over 4000 generations establishes a steady POF which already considers worst-case corners, namely, slow-n/slow-p, slow-n/fast-p, fast-n/slow-p, and fast-n/fast-p device models. The POF is shown in Fig. 4.5, with the fabricated solution highlighted.

4.3 Voltage-Combiner Biased OTA with Current Starving

The optimization of the amplifiers, in this chapter, is carried out using a 130 nm CMOS technology. In particular, the UMC 130 nm design kit, with standard-VT 3.3 V devices, is considered. For the implementation in a fully differential configuration, a common-mode feedback circuit is incorporated in the circuit, similar to the one presented in Fig. 4.3. The performance potential of the proposed amplifier is explored directly using a state-of-the-art analog IC design automation framework, AIDA-C [1]. In order to completely explore the trade-offs between the energy efficiency and the gain of the proposed amplifier, an IC sizing and optimization framework, AIDA-C [1], is used to automatically synthesize the circuit at sizing level.

The optimization method is carried out considering two objectives: the maximization of the low-frequency gain and the maximization of the FOM as in (2.14), where C_L is the load capacitance and I_{DD} is the current consumption, not considering

Table 4.2 VC biased OTA
with current starving
optimization performance and
functional constraints

Metrics	Specifications	Metrics	Specifications
GBW	$\geq$180 MHz	OVD	$\geq$0.07 V
Gain	$\geq$55 dB	DEL	$\geq$0.07 V
I_{DD}	$\leq$3 mA	AA	$\geq$1 μm^2
PM	$\geq$55°	WF	$\geq$0.72 μm
FOM	$\geq$850 MHz × pF/mA	VOS	$\leq$5 mV

external biasing sources, in this case I_{REF}. This resulted in several sizing solutions, constituting a POF, described further. Performance and functional constraints are taken into consideration in the method. The functional constraints are the overdrive voltages of the devices and their minimal saturation margins, while the performance constraints are related to the performance metrics of the proposed OTA, e.g., low-frequency gain, phase margin, and GBW. The devices are set to operate in moderate inversion, with a comfortable saturation margin to minimize the distortion. In this case, the N0, N1, N4, and N5 NMOS devices suffer from body effect, i.e., vsb > 0 V, since the bulks are connected to the ground. The constraints are shown in Table 4.2. The GBW is evaluated considering a 6 pF capacitive load. A maximum value of 5 mV, i.e., 0.15% of V_{DD}, for the offset voltage is set. In order to avoid fabrication process and mismatch impact, all active areas and width-*per*-finger are accounted. Again, the optimized sizing variables are the channel widths, channel lengths, and the number of fingers. A CMFB circuit, similar to the one shown in Fig. 4.3, is also considered in the optimization process, with functional specifications according. This circuit compares the output DC voltage with half of V_{DD}, in this case 1.65 V, setting the output on approximately 1.65 V through signal reinjection in the cmfb node. The optimization is carried out considering typical condition models of the UMC 130 nm kit. A 27 °C simulation temperature, a 3.3 V biasing voltage, and a 1.65 V common-mode voltage are considered. The optimization of this circuit, in the described context, takes approximately 3 h in an i7-3770 Intel© CPU with 16 GB RAM. A population of 128 elements evolves over 5000 generations, establishing a steady POF. The POF is shown in Fig. 4.6, with the fabricated solution highlighted in purple. The output of AIDA-C shown in Fig. 4.6 highlights the fabricated solution.

4.4 Folded Voltage-Combiner Biased OTA

The optimization of the amplifiers, in this chapter, is carried out using a 130 nm technology. In particular, the UMC 130 nm design kit, with standard-VT 1.2 V devices, is considered. In the case of the amplifier described in this section, no prototype fabrication is targeted; therefore, the work presented is only at proof-of-concept level. The implementation of a fully differential configuration of the amplifier described in this section demands the usage of a CMFB circuit. One possible implementation is presented in Fig. 4.7, which derives directly from the approach

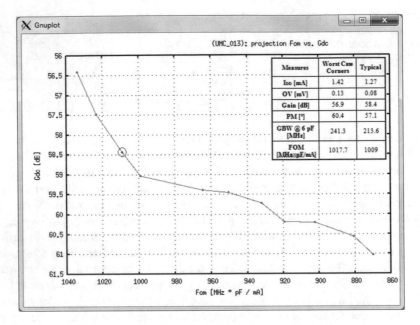

Fig. 4.6 VC biased OTA with current starving optimization POF considering typical conditions

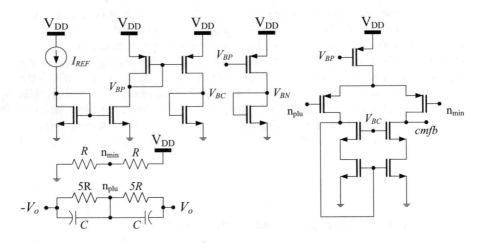

Fig. 4.7 Common-mode feedback circuit for the folded VC biased OTA

presented in Fig. 4.3, with the addition of a branch to bias the VC. As alternative, the implementation presented in Fig. 4.8 exempts the usage of an additional branch, when compared to the basic structure shown in Fig. 4.7. However, the latter pays a price in area due to the additional resistor. In fact, the sizing of the second circuit is easier, in the sense that V_{BN} is extracted directly from the I_{REF} branch. The results

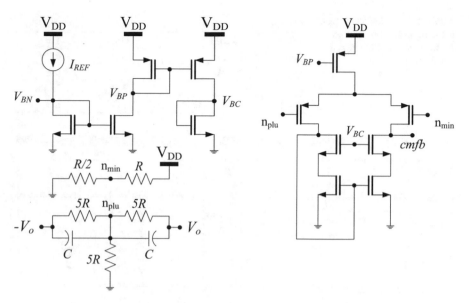

Fig. 4.8 Modified common-mode feedback circuit for the folded VC biased OTA

presented in this chapter, however, where achieved with the usage of an ideal and continuous-time CMFB, for proof-of-concept.

In order to evaluate the circuit performance and design trade-offs, the UMC 130 nm CMOS technology node is considered with standard-VT 1.2 V MOSFET devices, and two objectives are set in AIDA-C [1]: to maximize the FOM and to maximize the low-frequency gain. By maximizing the FOM, the power consumption is minimized, while the GBW is maximized, as in (2.14). In this case, it is used an ideal continuous-time CMFB circuit in the test bench of the circuit, for biasing the circuit and to settle its output voltage. In addition to the previously established objectives, performance and functional specifications are also considered in the optimization process. The performance specifications are $I_{DD} \leq 350$ µA, Gain $\geq$ 50 dB, GBW for 6 pF load ≥ 30 MHz, PM $\geq 60°$, and FOM ≥ 1000 MHz × pF/mA. The functional specifications of the devices are $V_{DS} -$ VDS$_{SAT} \geq 50$ mV and overdrive voltages ≥ 100 mV for all devices, enabling them to work on moderate/ strong inversion region, to avoid distortion. In this case, no body effect affects any device since all bulks are connected to the respective sources of the devices, i.e., the source voltage equals the substrate voltages in all devices. The specifications are shown in Table 4.3. The optimization variables are some of the channel widths, channel lengths, and number of fingers of the MOS devices.

The circuit optimization is executed for two different scenarios: first for typical working conditions and second for more stringent constraints; they are considered corner analysis conditions. For the typical analysis conditions, a temperature of 27 °C, a fixed supply voltage of 1.2 V, and the typical models of the devices are considered. For the corner analyses conditions, the same temperature and supply

Table 4.3 Folded VC biased OTA optimization performance and functional constraints

Metrics	Specifications	Metrics	Specifications
GBW	$\geq$30 MHz	OVD	$\geq$0.1 V
Gain	$\geq$50 dB	DEL	$\geq$0.05 V
I_{DD}	$\leq$350 µA	AA	$\geq$1 µm^2
PM	$\geq$60°	WF	$\geq$0.72 µm
FOM	$\geq$1000 MHz × pF/mA	VOS	$\leq$2 mV

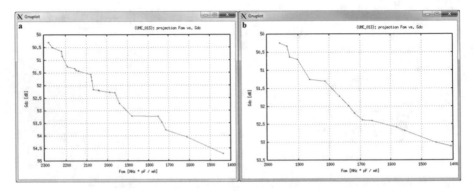

Fig. 4.9 Folded VC biased OTA POF: (**a**) typical conditions; (**b**) process corner conditions

Table 4.4 Simulated folded VC biased OTA POF extreme elements considering typical conditions

Solutions	FOM [MHz × pF/mA]	Gain [dB]	I_{DD} [mA]	PM [°]	GBW [MHz]	OS [V]
▲FOM	2278.5	50.3	0.218	61.0	82.79	0.693
▲Gain	1435.3	54.7	0.215	63.5	51.32	0.790

voltage values are taken into consideration, as well as the following models: slow NMOS and slow PMOS (SS), slow NMOS and fast PMOS (SNFP), fast NMOS and slow PMOS (FNSP), and finally, fast NMOS and fast PMOS (FF). In typical conditions, it is considered a 2000 generations evolution within a population of 128 elements. As for the process variations corner analyses scenarios, it is considered a 1000 generations evolution within a population of 128 elements. The complete design process takes approximately 6 h in a 2 GB RAM virtual machine running on an i7-2600 Intel® CPU. The Pareto front of the solutions in typical and corner conditions is shown in Fig. 4.9. The performance values for the extreme elements, i.e., the solutions with highest FOM and highest gain, of the previously quoted POFs are detailed in Tables 4.4, 4.5, and 4.6. Throughout the optimization process, the circuit is simulated using Mentor's Eldo® electrical simulator. It is possible to observe that, despite the degradation of the performance due to process variations, the obtained values corresponding to an extremely robust solution are still very high, namely, in terms of FOM, low-frequency gain, and current consumption [7]. Moreover, the SS case reveals to be the worst situation in terms of the overall performance, mainly due to the increase of all threshold voltages, translated in a lower GBW in practical terms.

Table 4.5 Simulated folded VC biased OTA POF highest FOM element considering corner conditions

▲FOM	Typical	SS	SNFP	FNSP	FF
FOM	2136.3	1959.3	2173.4	2097.5	2223.1
GBW [MHz]	72.07	62.1	71.29	71.47	78.84
Gain [dB]	50.36	50.42	50.67	50.45	50.25
I_{DD} [mA]	0.202	0.19	0.197	0.204	0.213
PM [°]	61.7	64.6	61.7	61.8	60.4
OS [V]	0.616	0.624	0.637	0.604	0.608

Table 4.6 Simulated folded VC biased OTA POF highest gain element considering corner conditions

▲Gain	Typical	SS	SNFP	FNSP	FF
FOM	1538.9	1405.9	1550.7	1528.2	1616.3
GBW [MHz]	50.28	43.5	49.23	50.47	54.92
Gain [dB]	53.12	53.2	53.5	53.3	53.1
I_{DD} [mA]	0.196	0.19	0.191	0.198	0.204
PM [°]	65.4	67.4	65.5	65	64.4
OS [V]	0.655	0.663	0.677	0.643	0.648

In order to verify the lowest biasing voltage achievable by this topology, additional optimizations are carried out, for the same specification of Table 4.3 with slight adjustments in the overdrive voltages, e.g., lowering them to 50 mV. The results of optimization processes, considering supply voltages from 1.1 V to 0.9 V, are shown in Fig. 4.10. It is impossible to obtain solutions operating below 0.9 V with reasonable overdrive and saturation margins, i.e., more than 50 mV. The results, for supply voltages from 1.2 V to 0.9 V, are shown in detail in Tables 4.7 and 4.8. In fact, for a supply source of 0.9 V, it is possible to obtain a solution with a FOM higher than 3000 MHz × pF/mA. The simulated AC response and step response of the highest FOM solutions for lower supply voltages are shown in Figs. 4.11, 4.12, and 4.13. Moreover, a modest degradation of results, e.g., a gain drop from 2 to 3 dB, and a maximum drop of 15% of the FOM are expected with the implementation of a CMFB as shown in Fig. 4.7.

4.5 Dynamic Voltage-Combiner Biased OTA

This section presents the dynamic circuit also optimized using AIDA-C, in a continuous-time domain, targeting high gain and energy efficiency, by means of FOM as defined in (2.14), i.e., in a continuous-time domain context, still having a direct relation with the FOM given by (3.38).

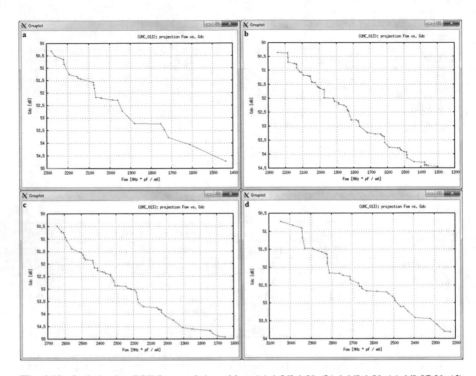

Fig. 4.10 Optimization POF for supply/overdrive: (**a**) 1.2/0.1 V; (**b**) 1.1/0.1 V; (**c**) 1/0.07 V; (**d**) 0.9/0.05 V

Table 4.7 Simulated POF performance metrics for lower supply voltages

Solutions	FOM [MHz × pF/mA]	Gain [dB]	I_{DD} [mA]	PM [°]	GBW [MHz]	OS [V]
▲FOM $w/V_{DD} = 1.2$ V	2278.5	50.3	0.218	61.0	82.79	0.693
▲FOM $w/V_{DD} = 1.1$ V	2245.8	50.4	0.188	60.0	70.22	0.739
▲FOM $w/V_{DD} = 1.0$ V	2645.1	50.5	0.139	60.9	61.40	0.677
▲FOM $w/V_{DD} = 0.9$ V	3043.0	50.7	0.110	60.1	55.93	0.621

Table 4.8 Simulated POF performance metrics for lower supply voltages: rise and fall time to input step

Solutions	Rise time [ns]	Fall time [ns]
▲FOM $w/V_{DD} = 1.2$ V	11.5	11.4
▲FOM $w/V_{DD} = 1.1$ V	16.1	16.0
▲FOM $w/V_{DD} = 1.0$ V	20.0	23.0
▲FOM $w/V_{DD} = 0.9$ V	24.1	29.9

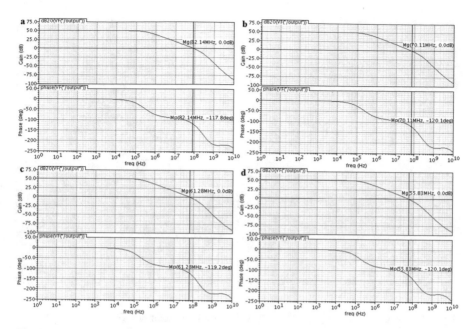

Fig. 4.11 Simulated AC magnitude and phase for the highest FOM solutions: (**a**) 1.2 V; (**b**) 1.1 V; (**c**) 1 V; (**d**) 0.9 V

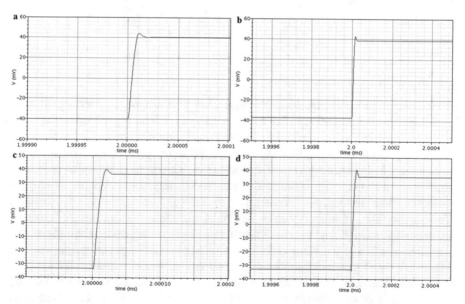

Fig. 4.12 Simulated step response (rise) for the highest FOM solutions: (**a**) 1.2 V; (**b**) 1.1 V; (**c**) 1 V; (**d**) 0.9 V

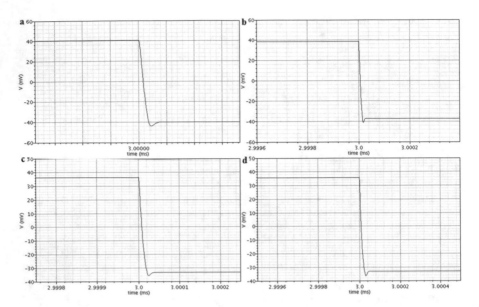

Fig. 4.13 Simulated step response (fall) for the highest FOM solutions: (**a**) 1.2 V; (**b**) 1.1 V; (**c**) 1 V; (**d**) 0.9 V

Table 4.9 Dynamic VC biased OTA optimization specifications

Metrics	Specifications	Metrics	Specifications
GBW	$\geq$50 MHz	OVD	$\geq$0.05 V
Gain	$\geq$60 dB	DEL	$\geq$0.05 V
I_{DD}	$\leq$2 mA	AA	$\geq$1 μm^2
PM	$\geq$40°	WF	$\geq$0.72 μm

In addition to the optimization objectives, key performance and functional specifications are also considered. The functional specifications are related to the overdrive voltages of the devices and to their minimal saturation margins. On the other hand, the performance specifications are related with the performance metrics of the circuit, e.g., low-frequency gain, PM, and GBW. The devices are set to operate in moderate inversion, with an adequate saturation margin to work as linearly as possible when needed, i.e., during $\emptyset_2$. A set of specifications are shown in Table 4.9. The minimum value of FOM for feasible solutions is set at 1000 MHz × pF/mA, and the GBW is evaluated considering a 6 pF load. Moreover, a minimum value of 3 mV for the offset voltage is set (2.5% of the biasing voltage, i.e., 1.2 V), and to avoid fabrication process and mismatch impact, all the active areas for all the devices and width-*per*-finger values are taken into account. Finally, the maximum spectral density of 30 nV/√Hz at GBW is set. In this case there is no body effect in any of the devices of amplifier, since the circuit is not fabricated, due to the difficult measurement as a stand-alone circuit. In order to test this circuit with reliability of the results, the latter has to be embedded either in an SC filter or in an ADC, which is considered outside of the scope of this work. The optimization variables are the channel widths, channel lengths, and the number of fingers of the devices. An ideal continuous-time CMFB circuit is employed.

The optimization is carried out considering corner conditions: slow-n/slow-p, slow-n/fast-p, fast-n/slow-p, fast-n/fast-p device models of the UMC 130 nm technology design kit. A temperature of 27 °C, a supply voltage at 1.2 V, and a 0.6 V common-mode voltage are considered. The optimization process for this circuit takes approximately 4 h in an i7-3770 Intel© CPU with 16 GB RAM. The sizing of the SC biasing branches relies on the static current consumption of active current sources, using (3.37), where R_{EQ} is the equivalent resistance of an active current source for a 100 MHz sampling clock frequency. The capacitor values are $C_1 = 216$ f. and $C_2 = 5.6$ pF. The AC response, shown in Fig. 4.14, during $\emptyset_2$ depicts a gain of 60.9 dB with a phase margin of 44.4°. During the reset, the quiescent current consumption is only 415 µA, as shown in Fig. 4.15, while in the amplification phase, the current consumption achieves the early peak value of 22.5 mA. The overall DC equivalent power, i.e., root mean square, consumption in each clock cycle, for a 1.2 V source, is 0.828 mW. A FOM of 1349 MHz × pF/mA$_{rms}$ is achieved. The current reference I_{REF} is not considered. The simulated noise response, shown in Fig. 4.16, depicts a 13 µV$_{rms}$ noise with a spectral density of 18.89 nV/√Hz at 155.1 MHz. The augmentation of noise, when compared to the static biasing, is related to the smaller current employed to bias the differential pair, in the end of the amplification phase. For proof of concept, a flipped-around S/H test bench for time-domain analyses is used. Coherent sampling is considered, using a Hamming truncation window and a 100 MHz clock frequency with 4096 samples,

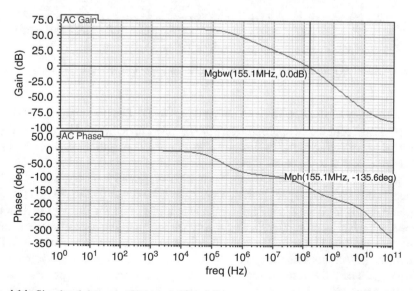

Fig. 4.14 Simulated dynamic VC biased OTA AC response

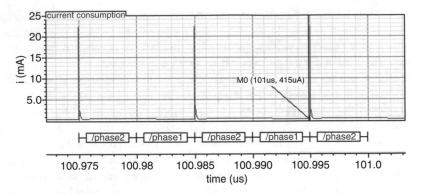

Fig. 4.15 Simulated dynamic VC biased OTA current consumption during two clock cycles

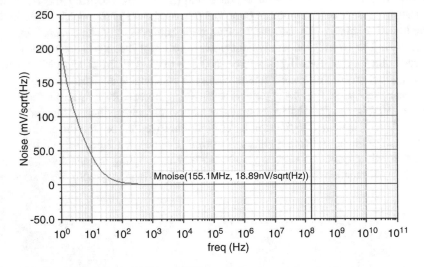

Fig. 4.16 Simulated dynamic VC biased OTA equivalent input-referred noise response

i.e., an accuracy of 12 bit, with no clock overlap. The DFT of the output signal shows a fundamental component approximately 68 dB better than the third harmonic which has a power of -89.35 dB, as illustrated in Fig. 4.17. A settling time of 21 ns is achieved for a 400 mV$_{pp}$ input step.

The design of a single-ended version is also presented with optimized performance. In this case, the capacitor values are $C_1 = 64.7$ f. and $C_2 = 1.79$ pF. The AC response during $\emptyset_2$ depicts a gain of 61 dB with a PM of 44.8°. During the reset the quiescent current consumption is 54.54 μA, while in $\emptyset_2$ the current consumption achieves a peak of 11.2 mA. The current reference I_{REF} is not considered. The DC equivalent power consumption, i.e., root mean square, in each clock cycle is 282 μW$_{rms}$. The noise response shows a 47.96 μV$_{rms}$ noise with a spectral density of approximately 82.14 nV/√Hz extracted at 51.73 MHz. A FOM of

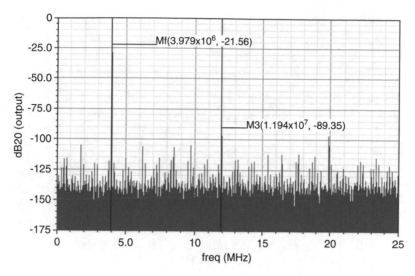

Fig. 4.17 Simulated dynamic VC biased OTA DFT response to a 4 MHz and 100 mV$_{pp}$ sinusoidal input

1320.8 MHz × pF/mA$_{rms}$ is achieved. When compared to the fully differential implementation, the single-ended version presents a higher noise response, which is, in part, related to the smaller current consumption of this version, since noise contribution and the drained current are inversely related. For proof-of-concept, a flipped-around S/H test bench for time-domain analyses is used. Coherent sampling is considered also in this case, using a Hamming window and a 100 MHz clock frequency with 4096 samples, i.e., 12 bit accuracy. The DFT of the output shows a fundamental component 70 dB better than the third harmonic, which has a power of −92.66 dB, as illustrated in Fig. 4.18. A settling time of 59 ns is achieved for a 0.4 V$_{pp}$ input step. The key performance parameters of both implementations are summarized in Table 4.10. Comparing with the two-stage OTA in [8], it is verifiable that the proposed solution has a higher FOM, with a lower current consumption.

4.6 EDA Techniques in Organic TFT Technologies

This section discusses and demonstrates the usage of electronic design automation techniques in organic thin-film transistor (OTFT)-based circuits. Particularly, this section presents the design and optimization of the VC biased OTA and the folded VC biased OTA, in single-ended configurations, using a top-gated carbon nanotube OTFT technology. The circuits are automatically designed and optimized at sizing level, using AIDA-C, in typical and in corner conditions, and considering a yield-aware sizing approach. A new figure-of-merit, more suitable to evaluate OTAs in this technology, is here proposed and justified. Simulation results show that DC gain

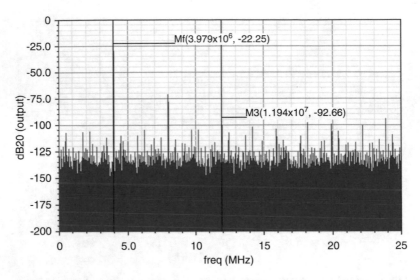

Fig. 4.18 Simulated single-ended dynamic VC biased OTA DFT response to a 4 MHz and 100 mVpp sinusoidal input

Table 4.10 Simulated key performance parameters of the dynamic VC biased OTA

Metrics	Proposed OTA (Single-ended)	Proposed OTA (Fully differential)
Gain [dB]	61	60.9
Phase margin [°]	44.8	44.4
GBW at 6 pF [MHz]	51.73	155.1
VOS [mV]	5.7	1.6
OS [V]	0.736	0.692
RMS current [mA$_{rms}$]	0.235	0.69
Noise [μV$_{rms}$]	47.96	11.03
Noise density [nV/√Hz]	82.14	18.89
FOM [MHz × pF/mA$_{rms}$]	1320.8	1349

values above 40 dB and FOM values beyond the state of the art in OTFT-based circuits can be achieved, thus contributing to advances in the context of OTFT-based circuitry and systems.

4.6.1 Context of Organic Technology

The integration of complete electronic systems in common environments, e.g., in flexible substrates such as clothes, plastic, and metal foils, will unquestionably improve security, convenience, and safety in everyday life. The OTFT technology has greatly expanded in recent years and presents itself as the most promising technology for these systems [9]. These systems require important features, such as mechanical flexibility and large-area/low-cost integration, which are not met by the existing silicon-based technologies, mainly due to their rigid structure and high fabrication temperatures [10]. One possible alternative would be thin chip technologies [11]. and substrate transfer from silicon-on-insulator wafers. These processes, however, require techniques that are currently not suitable to large-area uses. They are complex and costly [9].

OTFTs, which are composed by polymers, allow low-temperature processing. Thus, electronic organic devices can be realized on multiple substrates, e.g., papers sheets, and clothing. Being flexible and able to cover large areas, organic electronic circuits could be integrated in consumer goods and on ambient surfaces. In addition, the capability of printing organic transistors also feeds the perspective of low-cost fabrication. Moreover, some organic materials in solution can be deposited and patterned with conventional printing processes. However, numerous intrinsic and extrinsic characteristics of organic electronics technologies limit the complexity of the circuits and their performance levels. Organic electronics generally aim at applications that do not need high performance. Thus, the state-of-the-art of analog and mixed-signal circuits implemented with OTFTs is still in a pioneering stage. In fact, very few attempts have been made on printable, low-voltage organic transistors. Particularly, there are very few reported single-stage amplifiers in complementary organic technologies [12–16].

Another important factor that limits the complexity of organic circuits is the high variability of OTFTs [17]. This results from the use of low-temperature and low-cost techniques, which permit limited control on the process. Also, the typical trade-offs in the performance of the circuits, e.g., gain-bandwidth product versus power consumption reduction and low-frequency gain versus energy-efficiency, are also present in OTFT-based analog circuits. Nevertheless, the capability of integrating on the same substrate in all the functionalities of a complete system is of major importance, to fulfill the nowadays trend to incorporate complete systems in longer-lasting battery-powered equipment and completely justify an effort in this area. In fact, integrating complete systems in battery-powered equipment is directly dependent on having low-power circuits. In turn, reducing power in analog circuitry is mainly related with the energy efficiency of the amplifiers used in the different building blocks of the processing chains.

Considering the analog circuitry trade-offs, the influence of voltage and temperature corner variations, plus the variability in fabrication, the usage of electronic design automation techniques with embedded yield estimation in the sizing and optimization of analog circuits using this technology is clearly justified.

This work addresses these matters, by presenting an automatic synthesis procedure at sizing level, using an organic process design kit (OPDK) [18]. This is a software package, based on silicon transistors, which uses real properties of the OTFTs. Finally, an in-house-developed yield estimation procedure is embedded in the automatic synthesis process, which expectantly improves the fabrication assurance and reliability.

4.6.2   Optimization Setup

In order to evaluate the circuit performance and design trade-offs, the OPDK is considered, with carbon nanotube (CNT) devices for proof of concept, and two contradictory objectives are set: the maximization of the low-frequency gain and the maximization of the FOM, i.e., the energy efficiency. Since the OTFT devices are typically slower than the silicon-based technologies, i.e., the amplifiers have inferior GBW, in this work a change to the silicon technology FOM given by (2.14) is proposed. For OTFT-based amplifiers, the FOM described in (4.4) is proposed, where C_L is the load capacitance and I_{DD} is the current consumption, without considering external biasing sources. The GBW is now evaluated in kilohertz.

$$
\text{FOM} = \frac{\text{GBW} \times C_L}{I_{DD}} \left[\frac{\text{kHz} \times \text{pF}}{\text{mA}} \right] \tag{4.4}
$$

In addition to the objectives established, performance and functional specifications are also considered. The functional specifications are related to the overdrive voltages of the devices and to their saturation margins, while the performance specifications are related with the circuit performance metrics, e.g., low-frequency gain, phase margin, or gain-bandwidth product. All the specifications for both cases are shown in Table 4.11. The minimum value of FOM set for feasible solutions is set at 100 kHz × pF/mA, and GBW is measured for a 37 pF load. The voltage offset is limited to 10 mV. The VC biased OTA and the folded VC biased OTA are supplied by a 20 V and a 10 V voltage sources, correspondingly (Table 4.12).

Considering the relatively high threshold voltages of the devices, e.g., 800 mV for the NMOS devices and 400 mV for the PMOS devices, setting overdrive voltages between 100 and 200 mV enables the devices to operate in moderate inversion, enhancing the gain values of the OTAs. The optimization variables are the length and the width of the devices. The ranges of the design variables are shown in Tables 4.13 and 4.14.

The optimization is carried out for two scenarios: first for typical working conditions and afterward for corner working conditions. For the typical analysis conditions, it is considered a temperature of 27 °C and fixed biasing voltages. For the

Table 4.11 Performance and functional specifications for the VC biased OTA

Metrics	Specifications	Metrics	Specifications
GBW	≥1 kHz	OVD	≥0.2 V
Gain	≥20 dB	DEL	≥0.2 V
PM	≥45°	I_{DD}	≤10 mA

Table 4.12 Performance and functional specifications for the folded VC biased OTA

Metrics	Specifications	Metrics	Specifications
GBW	≥1 kHz	OVD	≥0.1 V
Gain	≥20 dB	DEL	≥0.1 V
PM	≥45°	I_{DD}	≤10 mA

Table 4.13 VC biased OTA optimization variables

Variables	Minimum	Grid unit	Maximum
Channel lengths	40 μm	1 μm	200 μm
Channel widths	50 μm	1 μm	1000 μm

Table 4.14 Folded VC biased OTA optimization variables

Variables	Minimum	Grid unit	Maximum
Channel lengths	25 μm	1 μm	200 μm
Channel widths	50 μm	1 μm	1000 μm

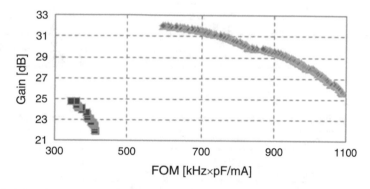

Fig. 4.19 POFs for the VC biased OTA: ● typical; ● corners

corner conditions, a variation of 10% is applied in the biasing voltage values and a variation of 7 °C in terms of temperature, in this way making four cases: HVHT, LVHT, HVLT, and LVLT. SS, FF SNFP, and FNSP are not yet available in the used OPDK. The results are presented in the next section. In the next paragraphs, the results of a yield-aware optimization, sighting the fabrication in a Fujifilm Dimatix DMP-2831 inkjet printer, are shown.

4.6.3   Optimization Results

In both typical conditions and voltage and temperature corner scenarios, it is considered a 12,000 generations evolution in populations of 128 elements each. The whole design process took approximately 12 h in an i7-3770 Intel© CPU with 16 GB of RAM. The POFs of the typical and corners, i.e., the optimization results, for the VC biased OTA are illustrated in Fig. 4.19. The highest FOM and gain solutions are presented in Table 4.15 in typical conditions. The highest FOM solutions in corner conditions are shown in Table 4.16, while the highest gain solutions in corner conditions are shown in Table 4.17. Throughout the design process, the circuits are simulated using Mentor's Eldo®.

Table 4.15 Voltage-combiner biased OTA extreme solutions in typical conditions

Objectives	FOM [kHz × pF/mA]	GBW [kHz]	Gain [dB]	I_{DD} [mA]	PM [°]
▲FOM	1089.9	121.1	25.7	4.1	47.0
▲Gain	594.9	126.4	32.0	7.9	47.0

Table 4.16 Voltage-combiner biased OTA highest FOM solution in corner conditions

▲FOM	Typical	HVHT	LVHT	HVLT	LVLT
FOM	415.8	417.4	412.9	417.4	412.9
GBW	83.4	89.0	87.9	79.3	78.3
Gain	22.0	22.0	21.9	22.0	21.9
I_{DD}	7.4	7.9	7.9	7.0	7.0
PM	47.2	47.3	47.0	47.3	47.0

Table 4.17 Voltage-combiner biased OTA highest gain solution in corner conditions

▲Gain	Typical	HVHT	LVHT	HVLT	LVLT
FOM	347.5	348.9	345.1	348.9	345.1
GBW	74.4	79.4	70.8	70.7	69.8
Gain	24.9	25.0	24.9	25.0	24.9
I_{DD}	7.9	8.4	8.4	7.5	7.5
PM	47.3	47.4	47.0	47.4	47.0

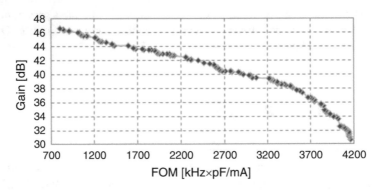

Fig. 4.20 POF for the folded VC biased OTA

The POF for typical conditions of the optimized folded VC biased OTA is shown in Fig. 4.20. The solutions with highest FOM and highest gain of the folded VC biased OTA are detailed in Tables 4.18 and 4.19. Simulation results, following the same approach as in the case of the VC biased OTA, are shown in Figs. 4.21, 4.22, and 4.23.

Table 4.18 Folded VC biased OTA extreme solutions in typical conditions

Objectives	FOM [kHz × pF/mA]	GBW [kHz]	Gain [dB]	I_{DD} [mA]	PM [°]
▲FOM	4166.8	122.2	30.7	1.1	47.0
▲Gain	802.1	96.5	46.6	4.5	47.2

Table 4.19 Yield-aware best solutions

Yield [%]	FOM [kHz × pF/mA]	GBW [kHz]	Gain [dB]	I_{DD} [mA]	PM [°]
72.3	465.6	111.5	27.4	8.86	49.5
51.3	1078.2	125.0	26.8	4.29	47.0
67.3	588.0	120.2	32.7	7.56	47.0
• 65.0	919.7	116.2	30.1	4.67	47.2

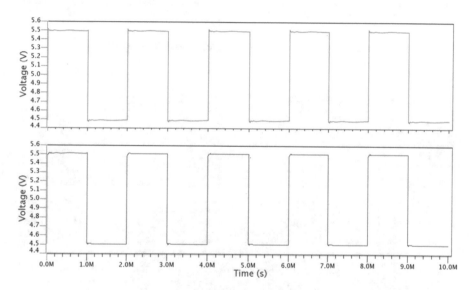

Fig. 4.21 Step response in unitary gain of the folded VC biased OTA: ▲FOM; ▲gain

In typical conditions, FOMs above 1000 kHz × pF/mA and 4000 kHz × pF/mA can be achieved, respectively, in the case of the VC biased OTA and in the case of the folded VC biased OTA. Comparing to [13], higher gain and GBW values are obtained. Despite the result degradation due to voltage and temperature variations, the obtained results are still very high in terms of FOM and gain, and the consumption is still low, according to the state of the art. The LVLT case reveals to be the worst situation in terms of overall performance, mainly due to the lowering of biasing voltage, i.e., typical minus 10%.

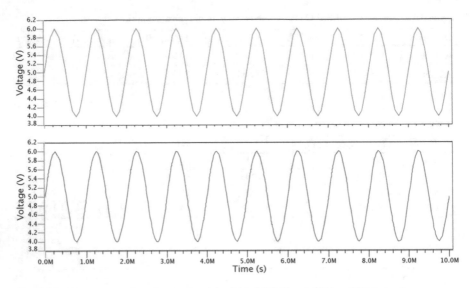

Fig. 4.22 Sine response in unitary gain of the folded VC biased OTA: ▲FOM; ▲gain

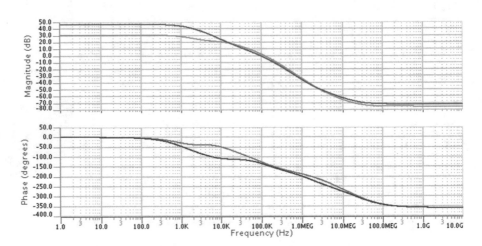

Fig. 4.23 AC response of the folded VC biased OTA: ▲ FOM magnitude response, ▲ FOM phase response, ▲ gain magnitude response, ▲ gain phase response

4.6.4 Yield Optimization Process

Considering the production of the organic single-ended VC biased OTA circuit, in a Fujifilm Dimatix DMP-2831 inkjet printer, a new circuit sizing optimization run needs to be setup. This new run takes advantage of a feature recently added to the circuit sizing optimization tool, the in-loop Monte Carlo-based yield optimization

[2]. This new feature allows performing the circuit sizing optimization taking under consideration the parametric circuit yield, which provides more robust solutions, and presents a yield estimation range for the different POF circuit solutions directly to the designer. The implemented Monte Carlo-based yield estimation technique selects only a small number of representative elements from the NSGA-II population to perform the Monte Carlo simulations, allowing the inclusion of this approach in the loop. The selection of the representative circuit solutions is performed by a modified K-Means clustering algorithm, where the solution with the best objective value *per* cluster is selected.

In order to perform Monte Carlo simulations, the standard silicon-based process design kits provide statistical variability device models. The lack of this type of models in the used OPDK represents an evident difficulty in the use of the Monte Carlo-based yield optimization process. In order to overcome this issue, the circuit netlist needs to be modified to include the process variability introduced by the Dimatix printer. The new netlist incorporates the reported drop size repeatability error into the widths and lengths of the transistors [19]. The second type or error introduced by the printing process, drop placement accuracy, is not considered because this error is only caused when the printer cartridge is replaced, and with a carefully planned printing strategy, it is possible to print each circuit layer with only one cartridge. The drop size repeatability printing error is modeled as Gaussian distribution centered in the transistor nominal widths and lengths, with standard deviations of 0.5%.

The yield optimization process performed 300 Monte Carlo iterations *per* generation and cluster, with a 99.7% confidence level. The number of clusters is defined by using the elbow criterion [20] and is set to six. The performance and functional specifications are the same as in the previous runs, as detailed in Table 4.11. Moreover, the design variables remain the same, as in Table 4.13.

The optimization objectives are the maximization of the yield, gain, and FOM. After some preliminary tests, it can be observed that the circuit yield is very sensitive to the circuit offset voltage; thus, it is increased. In fact, since the systematic offset can, in general, be less problematic once the VC biased OTA is operating in closed-circuit configuration, an optimization with less stringent offset voltage maximum values, i.e., in the order of hundreds of millivolts, in particular 0.1 V, can be setup. In these conditions, an optimization that considers the maximization of the yield, gain, and FOM is carried out, corresponding to a run of 1000 generations with a population of 256 individuals, with an approximated duration of 5 h in an i7-3770 Intel© CPU with 16 GB RAM.

The results for this optimization process are shown in Fig. 4.24. Moreover, the highlighted solution and the solutions that correspond to the best optimization values *per* objective are presented and detailed in Table 4.19. For proof-of-concept, an AC and a transient analysis are carried out using Mentor's Eldo®, for the intermediate solution highlighted in Fig. 4.24. The results are shown in Figs. 4.25 and 4.26. The selected solution as also tested in 3000 Monte Carlo iterations. The Monte Carlo run, depicted in Fig. 4.27, reveals a final yield of 65% with high precision results.

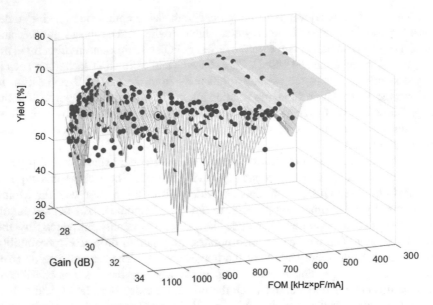

Fig. 4.24 Pareto surface for the yield-aware tridimensional optimization process

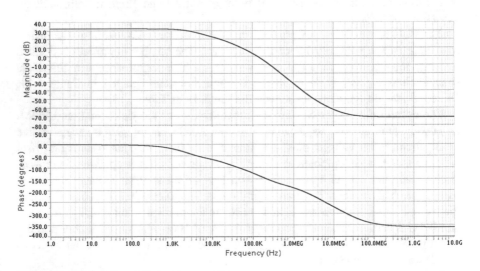

Fig. 4.25 AC response of the VC biased OTA

The obtained solutions achieve a maximum yield up to 72.3%; in the FOM objective, it is possible to obtain values above 1000 kHz × pF/mA, and on the opposite extreme of the POF, a solution with a gain of 32.7 dB can be observed.

A comparison with recently published state-of-the-art amplifiers using OTFT is shown in Table 4.20, where it is possible to acknowledge the performance capabilities of the presented designs, namely, in terms of GBW and FOM, as well as the

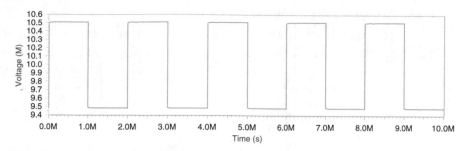

Fig. 4.26 Step response in unitary gain of the VC biased OTA

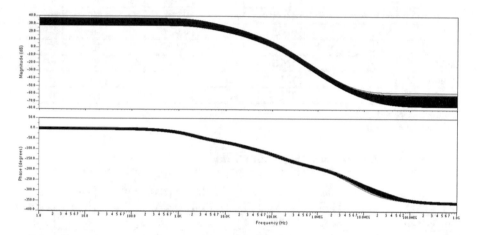

Fig. 4.27 Monte Carlo AC response of the VC biased OTA

potential of the optimization procedure. The presented solutions trade consumption for GBW, keeping high gains, for high precision applications.

4.7 Summary

This chapter presents the optimized results of a new family of single-stage OTAs, biased by voltage combiners, with high gain and energy efficiency. In this new family, the traditional current source that typically bias the differential pair of a symmetrical CMOS OTA is replaced by several configurations of voltage combiners, i.e., in a cross-coupled configuration, in a folded configuration to specifically tackle low voltage supplies, and finally in a dynamic configuration to be implemented in high-performance ADCs and SC filters. Finally, an implementation of two topologies using an organic technology is addressed, with promising results. A summary of the works proposed and respective implementations is presented in Table 4.21, which comprises results at simulation level, before and after optimization in AIDA-C. The potential of both the new family of CMOS amplifiers and the

Table 4.20 Comparison with state-of-the-art OTFT amplifiers

References	FOM [kHz × pF/ mA]	GBW [kHz]	Gain [dB]	I_{DD} [mA]	PM [°]	Bias [V]	Load [pF]
[13]	–	–	27/22	0.001	–	40	–
[14]	–	3	10	0.005	90	10	–
[21]	–	2	28	0.015	–	2	–
[15]	–	0.055	27	–	–	60	–
[22]	–	0.0075	36	0.055	–	5	–
[23]	–	0.5	23	0.021	70	15	–
[24]	800	2	20	0.015	65	15	6
[16]	6923	1.5	40	0.013	58	50	60
[16]	643	0.075	51	0.007	70	50	60
[16]	5775	0.055	49	0.002	60	50	210
VC[a]	1090	121.1	26	4.1	47	20	37
VC[b]	594.9	126.4	32	7.9	47	20	37
FoldedVC[a]	4167	122.2	31	1.1	47	10	37
FoldedVC[b]	802	96.5	47	4.5	47	10	37

[a]Highest FOM solution
[b]Highest gain solution

Table 4.21 Summary table of all the proposed works

References	Gain [dB]	I_{DD} [mA]	PM [°]	GBW [MHz]	FOM [MHz × pF/ mA]	Load [pF]	Tech [nm]
VC biased OTA[a]	45.0	1.300	84.1	51.2	236.3	6	130
VC biased OTA[b]	46.4	1.07	58.4	169.5	950.5	6	130
VC biased OTA w/current Starving[a]	63.1	2.000	60.1	146.7	440.1	6	130
VC biased OTA w/current Starving[b]	58.4	1.270	57.1	213.6	1009.0	6	130
Folded VC biased OTA[a]	40.2	0.400	73.3	67.6	1014.0	6	130
Folded VC biased OTA[b]	54.7	0.215	63.5	51.3	1435.3	6	130
Folded VC biased OTA[b]	50.3	0.218	61.0	82.8	2278.5	6	130
Dynamic VC biased OTA SE[a]	50.1	0.251	46.6	33.1	792.0	6	130
Dynamic VC biased OTA FD[a]	54.2	0.494	52.6	85.5	1038.8	6	130
Dynamic VC biased OTA SE[b]	61	0.235	44.8	51.7	1320.8	6	130
Dynamic VC biased OTA FD[b]	60.9	0.69	44.4	155.1	1349.0	6	130

[a]Simulation results before optimization
[b]Simulation results after optimization

optimization framework is clearly demonstrated, since the values of FOM and gain are greatly enhanced after optimization: the FOM of the amplifiers are, in general, better than 1000 MHz $\times$ pF/mA, representing the accomplishment of the objective proposed in this work, i.e., surpassing the value of FOM that is possible to achieve with classic folded-cascode topologies.

References

1. N. Lourenço, et al., "AIDA: Robust Analog Circuit-Level Sizing and In-Circuit Layout Generation," in Integration, the VLSI Journal, 2016. DOI: https://doi.org/10.1016/j.vlsi.2016.04.009.
2. A. Canelas, et al., "Yield Optimization using K-Means Clustering Algorithm to Reduce Monte Carlo Simulations," in IEEE International Conference on Synthesis, Modeling, Analysis and Simulation Methods and Applications to Circuit Design (SMACD), Page(s): 1–4, Jun. 2016. DOI: https://doi.org/10.1109/SMACD.2016.7520729.
3. N. Lourenço, et al., "Floorplan-Aware Analog IC Sizing and Optimization based on Topological Constraints," in Integration, the VLSI Journal, Vol. 48, Page(s): 183–197, Elsevier, Jan. 2015. DOI: https://doi.org/10.1016/j.vlsi.2014.07.002.
4. R. Martins, et al., "LAYGEN-II Automatic Layout Generation of Analog Integrated Circuits," in IEEE Transactions on Computer-Aided Design of Integrated Circuits and Systems, Vol. 32, Issue 11, Page(s): 1641–1654, Nov. 2013. DOI: https://doi.org/10.1109/TCAD.2013.2269050.
5. B. Liu, et al., "Automated Design of Analog and High-frequency Circuits," in Fundamentals of Optimization Techniques in Analog IC Sizing, Page(s): 19–40, Springer, 2014. DOI: https://doi.org/10.1007/978-3-642-39162-0_2.
6. K. Deb, et al., "A fast and elitist multiobjective genetic algorithm: NSGA-II," in IEEE Transactions on Evolutionary Computation, Vol. 6, Issue 2, Page(s): 182–197, Aug. 2002. DOI: https://doi.org/10.1109/4235.996017.
7. R. Assaad, J. Silva-Martinez, "The Recycling Folded-cascode: A General Enhancement of the Folded-Cascode Amplifier," in IEEE Journal of Solid-State Circuits, Vol. 44, Issue 9, Page(s): 2535–2542, Sep. 2009. DOI: https://doi.org/10.1109/JSSC.2009.2024819.
8. A. Zadeh, "A 100MHz, 1.2V, ±1V Peak-to-Peak Output, Double-Bus Single Ended-to-Differential Switched Capacitor Amplifier for Multi-Column CMOS Image Sensors," in IEEE International New Circuits and Systems Conference (NEWCAS), Page(s): 1–4, Jun. 2016. DOI: https://doi.org/10.1109/NEWCAS.2016.7604739.
9. S. Abdinia, et al., "Design of Organic Complementary Circuits and Systems on Foil," Springer, 2015. ISBN: 978-3-319-21187-9.
10. H. Klauk, "Organic Thin-Film Transistors", in Chemical Society Reviews, Vol. 39, Page(s): 2643–2666, 2010. DOI: https://doi.org/10.1039/B909902F.
11. J. Burghartz, et al., "Ultra-Thin Chips and Related Applications, a New Paradigm in Silicon Technology," in Proceedings of the IEEE European Solid-State Circuits Conference, Page(s): 28–35, 2009. DOI: https://doi.org/10.1109/ESSDERC.2009.5331441.
12. V. Vaidya, et al., "An Organic Complementary Differential Amplifier for Flexible AMOLED Applications," in Proceedings of IEEE International Symposium on Circuits and Systems (ISCAS), Page(s): 3260–3263, 2010. DOI: https://doi.org/10.1109/ISCAS.2010.5537910.
13. M. Guerin, et al., "High-Gain Fully Printed Organic Complementary Circuits on Flexible Plastic Foils," in IEEE Transactions on Electron Devices, Vol. 58, Issue 10, Page(s): 3587–3593, Aug. 2011. DOI: https://doi.org/10.1109/TED.2011.2162071.

14. M. Torres-Miranda, et al., "High-Speed Plastic Integrated Circuits: Process, Design, Integration and Test," in IEEE Journal on Emerging and Selected Topics in Circuits and Systems, Vol. 7, Issue 1, Mar. 2016. DOI: https://doi.org/10.1109/JETCAS.2016.2611823.
15. J. Chang, et al., "Fully printed electronics on flexible substrates: High gain amplifiers and DAC," in Organic Electronics, Vol. 15, Issue 3, Page(s): 701–710, Mar. 2014. DOI: https://doi.org/10.1016/j.orgel.2013.12.027
16. G. Maiellaro, et al., "High-Gain Operational Transconductance Amplifiers in a Printed Complementary Organic TFT Technology on Flexible Foil," in IEEE Transactions on Circuits and Systems-I, Vol. 60, Issue 2, Page(s): 3117–3125, Dec. 2013. DOI: https://doi.org/10.1109/TCSI.2013.2255651.
17. H. Marien, et al., "Analog Techniques for Reliable Organic Circuit Design on Foil Applied to an 18 dB Single-Stage Differential Amplifier," Organic Electronics, Vol. 11, Issue 8, Page(s): 1357–1362, Aug. 2010. DOI: https://doi.org/10.1016/j.orgel.2010.05.006.
18. Wei Zhang, "OPDK User Manual," Release 2014.3, University of Minnesota, Mar. 2014.
19. Fujifilm Dimatix, "Materials Printer & Cartridge DMP-2800 Series Printer & DMC-11600 Series Cartridge FAQ report," Apr. 2016.
20. P. Bholowalia, et al., "EBK-Means: A Clustering Technique based on Elbow Method and K-Means in WSN," in International Journal of Computer Applications, Vol. 105, Issue 9, Page (s): 17–24, Nov. 2014. DOI: https://doi.org/10.5120/18405-9674.
21. H. Fuketa, et al., "1 µm-Thickness ultra-flexible and high electrode-density surface electromyogram measurement sheet with 2 V organic transistors for prosthetic hand control," in IEEE Transactions on Biomedical Circuits and Systems, Vol. 8, Issue 6, Page(s): 824–833, 2014. DOI: https://doi.org/10.1109/TBCAS.2014.2314135
22. I. Nausieda, et al., "Mixed-signal organic integrated circuits in a fully photolithographic dual threshold voltage technology," in IEEE Transactions on Electron Devices, Vol. 58, Issue 3, Page(s): 865–873, Mar. 2011. DOI: https://doi.org/10.1109/TED.2011.2105489.
23. H. Marien, et al., "A fully integrated sigma-delta ADC in organic thin-film transistor technology on flexible plastic foil," in IEEE Journal of Solid-State Circuits, Vol. 46, Issue 1, Page(s): 276–284, Jan. 2011. DOI: https://doi.org/10.1109/JSSC.2010.2073230.
24. H. Marien, et al., "Analog building blocks for organic smart sensor systems in organic thin-film transistor technology on flexible plastic foil," in IEEE Journal of Solid-State Circuits, Vol. 47, Issue 7, Page(s): 1712–1720, Jul. 2012. DOI: https://doi.org/10.1109/JSSC.2012.2191038.

Chapter 5
Integrated Prototypes and Experimental Evaluation

5.1 Test Printed Circuit Board

The VC-biased fully differential and the VC-biased fully differential amplifiers with current starving have both been fabricated and manually wire-bonded using a TPT HB10 semiautomatic bonding machine to a manually designed test printed circuit board (PCB). The PCB is completely designed in Eagle 6.1.0 framework [1]. The design of this PCB is also, in part, based on the original Texas Instruments test board THS4521/2/4EVM SOIC-8 package [2]. The electrical scheme comprises two transformers, one on the input and another on the output of the setup, to simplify the conversion from single-ended to differential terminations, two output buffers, one voltage-reference circuit to generate a bias voltage, and an input current/common-mode voltage generator. It is possible to avoid the usage of the output buffers if they are not needed, bypassing them directly. In fact, the designed test PCB is able to cover the key aspects of fully differential amplifier testing, i.e., static power consumption, rise time, fall time, slew rate, and frequency response (low-frequency gain, bandwidth and gain-bandwidth product, and noise response), for SOIC-8- packages and also typical UMC 130 nm prototype dies. The main advantage of this work is that the open-circuit gain can be obtained in a single closed-circuit configuration, using a common network analyzer. All supply nodes have capacitive connections to ground to filter undesired fluctuations. The layout of the PCB test board is designed using the two-layer PCB feature. For the purposes of this work, it is simply not found necessary to use more layers. The signal path is mainly written on the top layer, while supply connections are mainly on the bottom layer of the PCB. Ground is divided by both layers. This way, it is possible to test the Texas Instruments amplifier if soldered on the bottom layer, i.e., the PCB has a SOIC-8 footprint on the bottom layer. The signal is brought back to the top layer. Moreover, it is possible to solder a 1.5 × 1.5 mm square die on the top layer to test the prototypes. The PCB is finally a rectangle of 14.5 × 9.5 cm. The dies,

© Springer International Publishing AG, part of Springer Nature 2019
R. F. S. Póvoa et al., *A New Family of CMOS Cascode-Free Amplifiers with High Energy-Efficiency and Improved Gain*,
https://doi.org/10.1007/978-3-319-95207-9_5

corresponding to the circuits presented in Sects. 3.2/4.2 and 3.3/4.3, i.e., the VC-biased OTA and the VC-biased OTA with current starving, have both been fabricated as described further, attached to the PCB using Electrolube SMA10SL Glue Adhesive, and finally, the circuit is protected with Robnor EL420AR/NC/ 050TC Adhesive. The components with fine tolerance are manually soldered in the PCB with 0603 components, and measurements have been carried out as presented in the further sections.

5.2   Voltage-Combiner-Biased OTA

The experimental evaluation of the VC-biased OTA is presented in this section. This circuit has been designed at layout level, prototyped, manually wire-bonded to a test PCB, and experimentally evaluated. The layout core of the VC-biased OTA sizing solution highlighted in Fig. 4.5 is shown in Fig. 5.1, comprising an area of 0.027 mm^2, i.e., 126.9 µm × 210.0 µm. Electrostatic discharge (ESD) protections are implemented with technology design kit components, following the circuitry in Fig. 5.1. The device-sizing symmetry of the circuit is transported to the layout design, as shown in Fig. 5.1, where the feeding top PMOS are shown in particular and where the NMOS core of the amplifier is shown. Moreover, the feeding top PMOS are isolated through a dedicated guard ring. In order to reduce the effects of parasitic capacitances, the signal paths are designed using higher metal levels, i.e., metals 3, 4, 5, and 6. The power nets are laid out using lower

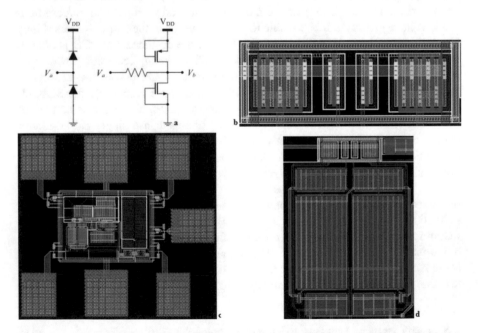

Fig. 5.1 VC-biased OTA layout: (**a**) ESD protections; (**b**) PMOS; (**c**) complete layout; (**d**) NMOS

Table 5.1 VC-biased OTA simulation results

Metrics	Worst-case corners	Typical	XRC extracted layout
DEL [mV]	78.91	☺	Not Available
OVD [mV]	51.03	☺	Not Available
WF [μm]	2.6	☺	☺
AA [μm^2]	1.2	☺	☺
I_{DD} [mA]	1.32	1.07	1.07
VOS [mV]	0.55	0.05	2.31
Gain [dB]	46.1	46.4	46.5
GBW at 6 pF [MHz]	206.8	169.5	163.6
PM [°]	63.4	58.4	65.8
FOM [MHz × pF/mA]	940.0	950.47	917.4

metal levels, i.e., metal 1 in the case of the ground routing and metal 2 in the case of the supply source routing. The branches are sized with 1 μm *per* 1 mA of direct current, preventing the impact of parasitic resistances in the routing. Moreover, given the fact that UMC 130 nm top two layers enhance the production cost, the two top layers, i.e., metal 7 and metal 8, remain unused. Typically, metals 7 and 8 are used for standard RF circuit implementations, e.g., inductors. Moreover, the layout core is circumscribed with an NWELL ring to avoid current leakage and for noise isolation purposes. The substrate and the NWELL have widespread connections to avoid latch-up events [3]. The simulated results before and after the layout design are summarized in Table 5.1.

The results presented in Table 5.1 are discussed further. The current consumption in Table 5.1 comprises the whole circuitry except I_{REF}. Particularly, the current consumption and the GBW, i.e., the important parameters for the energy-efficient FOM considered in (2.14), do not vary significantly from the typical conditions to the XRC extracted layout values of post-layout simulations. In terms of gain, a maximum variation of less than 1% is depicted, if typical and extracted results are concerned. In fact, the most important variations are reflected in the offset voltage. This is usually not problematic, since the OTA is used in closed-circuit configurations, and the offset voltage is simply compensated and cancelled out. In order to validate the obtained results, post-layout AC and transient analyses are carried out, in typical conditions. The analysis in the time domain is carried out in a closed circuit of unitary gain. The post-layout step response demonstrates some initial ringing, yet a stable response with a rise time of 23 ns and a fall time of 36 ns, with a single-ended OS of 2.464 V. The post-layout noise response depicts 5.8 nV/√Hz at GBW. Considering a center frequency of approximately 15 Hz, an integrated flicker noise value of 15.66 μV$_{rms}$ can be achieved. This is important, due to the large number of devices connected to the inputs of the amplifier. A thermal noise of approximately 80.9 μV$_{rms}$ is achieved at simulation level. A summary of state-of-the-art noise contributions in recent literature is presented in Table 5.2, where it is possible to verify that the proposed topology is in line with the state of the art. The post-layout PSRR depicts a value of approximately 109 dB, when 1 V of amplitude is delivered

Table 5.2 Comparison with state-of-the-art contributions in terms of noise parameters

Work	Center frequency	Bandwidth	Flicker	Thermal	Integrated noise
[5]	–	1 Hz–100 MHz	–	–	48.5 μV_{rms}
[6]	–	1 Hz–100 MHz	–	–	118 μV_{rms}
VC-biased OTA	15 Hz	1 Hz–170 MHz	15.7 μV_{rms}	80.9 μV_{rms}	96.6 μV_{rms}

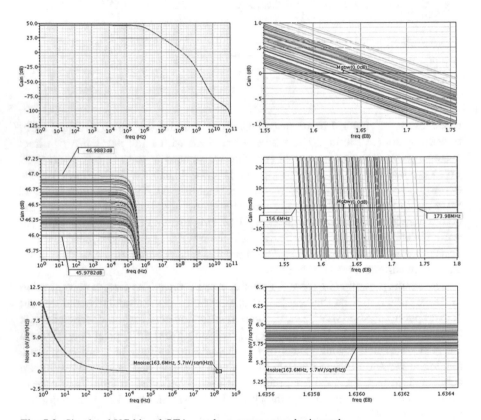

Fig. 5.2 Simulated VC-biased OTA post-layout process and mismatch

by the supply. The value is in line with prior artwork in [4], which depicts a PSRR of 74 dB. The common-mode gain of is simulated at post-layout conditions, and a value of less than −62 dB is achieved at baseband. By subtracting the AC common-mode gain to the AC gain response, a CMRR of more than 100 dB is estimated, which corresponds to more than three times the actual small-signal AC gain. The obtained value of CMRR is in line with [7], which shows that for most common applications, a value of CMRR stands 70 dB and 120 dB at low frequencies. The obtained value, however, goes beyond what is presented in [4], i.e., a CMRR of 62.3 dB. Moreover, a 3-σ Monte Carlo (MC) simulation with 100 runs is carried out to acknowledge the robustness of the circuit, and the results are shown in Fig. 5.2. The results depict an

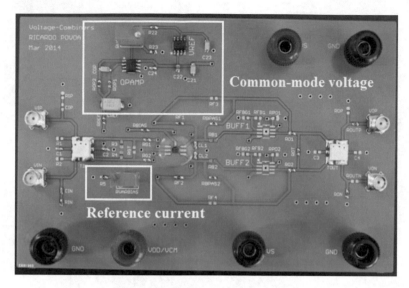

Fig. 5.3 Fabricated PCB with wire-bonded die and components for VC-biased OTA AC measurements

expected variation of 1.3% and 7.6% in the gain and GBW, correspondingly. The test PCB to experimentally evaluate the VC-biased OTA, with the die attached and all components soldered for AC measurements, is shown in Fig. 5.3.

In order to test the amplifier, the components are soldered to the board, according to the simplified circuit shown in Fig. 5.4 as described in [2, 8]. The current consumption cannot be directly measured, because the common-mode voltage generation circuitry has to be enabled, for the OTA to be operational. Therefore, the consumption of the circuit is measured indirectly by subtracting the consumption of the common-mode voltage circuitry from the overall current consumption of the PCB, resulting in 1 mA, which already includes all the power dissipation of the internal CMFB circuit. The common-mode voltage of the OTA is set to 1.66 V. In a closed circuit, the OTA sets an output DC voltage of 1.73 V, which means the systematic offset voltage is 70 mV. This aspect can be justified with fabrication deviations and parasitic resistive effects of the PCB. The step response, shown in Fig. 5.5, is analyzed in a unitary gain circuit, as in [2]. A settling time of 0.1 μs is measured, for a 100 kHz 400 mV input. A rise time of 21.95 ns and a fall time of 33.91 ns are also verifiable experimentally, as shown in Fig. 5.5.

The GBW and low-frequency gain of the circuit are measured using a unitary closed-circuit configuration, described in detail in [2] and [8]. In order to test the amplifier in the frequency domain, the reference of the network analyzer is put in the input of the amplifier. The signal is read in the positive output of the setup, with a Hewlett Packard 4195A network/spectrum analyzer and with a 0 dBm input signal. The AC response, with emphasis on the GBW, is shown in Fig. 5.6 depicting the value of approximately 170.6 MHz to a 6 pF load. A low-frequency gain around 47 dB is achieved. The better GBW obtained experimentally is mostly related to the

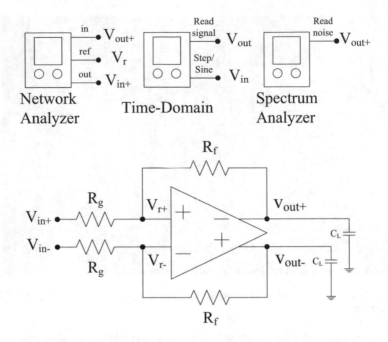

Fig. 5.4 VC-biased OTA experimental testing workbench

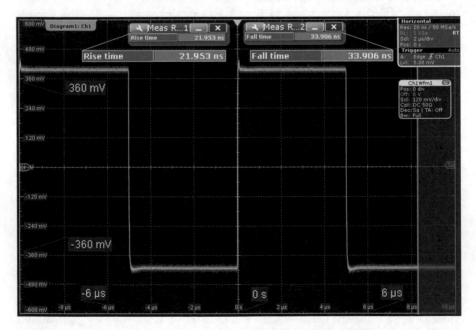

Fig. 5.5 Measured VC-biased OTA step response: rise time, 21.95 ns; fall time, 33.91 ns

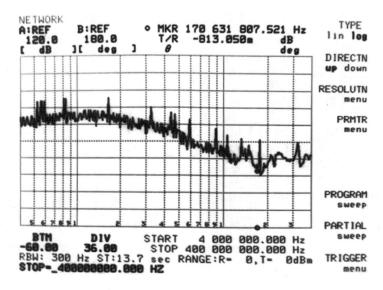

Fig. 5.6 Measured VC-biased OTA AC gain response

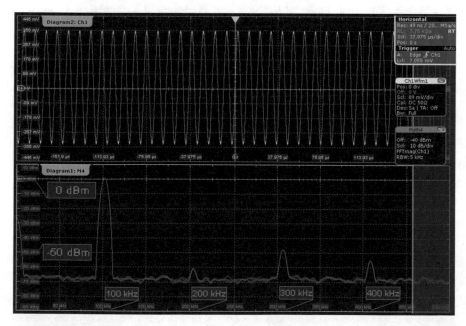

Fig. 5.7 Measured VC-biased OTA harmonic distortion analysis

real load capacitance of the test workbench, which proves to be slightly lower than the initial estimated value, i.e., 6 pF. In order to measure the distortion, a 100 mV and 100 kHz sinusoidal signal is set at the input of the OTA in a closed circuit with a gain of 3 V/V. The results are in Fig. 5.7, where it is possible to verify that the third

Table 5.3 VC-biased OTA compared with recently published single-stage amplifiers

References	Gain [dB]	I_{DD} [mA]	PM [°]	GBW [MHz]	FOM [MHz × pF/mA]	Load [pF]	Tech [nm]
[5]	60.9	0.800	71	134.2	939.4	5.6	180
[9]	54	1.000	62	965.0	965.0	1.0	180
[6]	85.6	1.000	67	987.0	987.0	1.0	180
[10]	72.0	0.652	70	159.0	536.5	2.2	180
[4]	84.0	0.085	81	12.5	146.0	1.0	180
[11]	94.9	3.670	82	414.0	225.0	2.0	130
[12]	91.5	5.000	62	714.5	1071.5	7.5	130
[13]	67.0	2.170	67	920.0	211.0	0.5	180
VC-biased OTA[a]	46.5	1.07	65.8	163.6	917.4	6	130
VC-biased OTA[b]	47	1	>55	170.6	1024	6	130

[a]Post-layout simulation results
[b]Experimental measurements

harmonic is 50 dBm better than the fundamental, confirming the approximately 50 dB gain. The input swing of the amplifier is measured by applying a sweep of sinusoidal signals of 100 kHz in a unitary closed circuit, from 100 mV to 2 V of amplitude. It can be verified that above 950 mV, the difference between the third harmonic and the fundamental components is clearly less than 50 dBm; thus, the input swing of the OTA is approximately 950 mV.

The proposed solution and a set of other recently published amplifiers are presented in Table 5.3. It is possible to verify that this circuit represents a relevant contribution to the state of the art in this field. This amplifier is, at this point, the third most energy-efficient solution, according to (2.14). As a matter of fact, when compared to the RFCA as presented in [5], the proposed solution has a considerably higher FOM.

5.3   Voltage-Combiner-Biased OTA with Current Starving

The experimental evaluation of the VC-biased OTA with current starving is presented in this section. This circuit has been designed at layout level; prototyped; manually wire-bonded to a test PCB, similar to the prototype presented in Sec. 5.1; and validated experimentally. The layout core of the VC-biased OTA with current starving sizing solution highlighted in Fig. 4.8 is shown in detail in Fig. 5.8, comprising an area of 0.037 mm^2, with several important details highlighted, as described further. The device sizing symmetry of the circuit is kept in the layout design, as is shown in Fig. 5.8, where the feeding top PMOS and the NMOS core of the amplifier are shown in particular. The feeding top PMOS are isolated through a dedicated guard ring. In order to reduce the effects of parasitic capacitances, the signal paths are laid

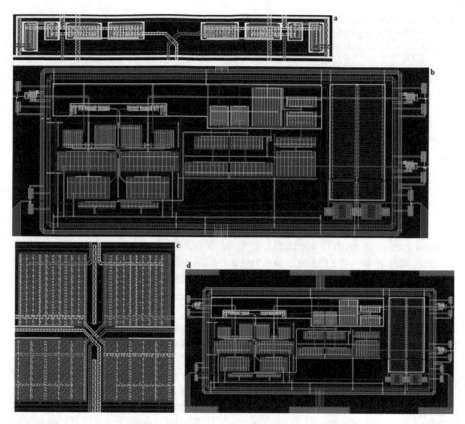

Fig. 5.8 VC-biased OTA with current starving layout core: (**a**) PMOS current mirrors; (**b**) layout core zoomed; (**c**) NMOS core with symmetry; (**d**) layout core with 129.0 μm × 290.0 μm = 0.037 mm^2

out using higher metal levels, i.e., metals 3, 4, 5, and 6. The power nets are designed using lower metal levels, i.e., metal 1 in the case of the ground routing and metal 2 in the case of the supply source. Both diode and active ESD protections are also implemented. The branches are sized with 2 μm *per* 1 mA of direct current to reduce the impact of parasitic resistances in the routing. The resistive effects have a higher impact as the complexity of the circuits increase, mainly in stability. Moreover, the output branches have a double routing with metal stripes in parallel, widely interconnected. Furthermore, the layout core is circumscribed with an NWELL ring to avoid current leakage and for noise isolation purposes.

The substrate and the NWELL have widespread connections to avoid latch up events. The complete layout, showing the pads for manual wire-bonding, is also shown in Fig. 5.8. No high-frequency effects, namely, transmission-line behavior, are present. Considering that $f \times \lambda = c$, where c is the light speed in vacuum, λ is the wavelength, and f is the wave frequency, it is possible to understand that for frequencies of 200 MHz, i.e., unit frequency of the amplifier, the wavelength is

Table 5.4 VC-biased OTA with current starving simulation results

Metrics	Worst-case corners	Typical	XRC extracted layout
DEL [mV]	☺	☺	Not Available
OVD [mV]	☺	☺	Not Available
WF [μm]	☺	☺	☺
AA [μm^2]	☺	☺	☺
I_{DD} [mA]	1.42	1.27	1.27
VOS [mV]	0.13	0.08	5.27
Gain [dB]	56.9	58.4	57.9
PM [°]	60.4	57.1	51.9
GBW at 6 pF [MHz]	241.3	213.6	195
FOM [MHz × pF/mA]	1017.7	1009	921.3

around 1 meter, which is not comparable with the full length of the circuit. Nevertheless, the devices are placed close to each other to reduce the impact of parasitic resistances. The simulated results before and after the layout design are summarized in Table 5.4. The current consumption in Table 5.4 comprises the whole circuitry except I_{REF}.

In order to verify and analyze the obtained results with higher detail, AC and transient analyses are made in typical conditions. First, a single-ended OS of 2.5 V is achieved with this topology. The response to a 1 kHz and 500 mV input step demonstrates a rise time of 16 ns and a fall time of 18 ns in a closed circuit of unitary gain. The noise response depicts a 3 nV/√Hz at the value of GBW. The post-layout PSRR depicts a value of approximately 119 dB, which stands in line with the state-of-the-art work in [4] that depicts a PSRR of approximately 74 dB. The common-mode gain of the amplifier is simulated at post-layout conditions, and a value of less than −46 dB is achieved at baseband. By subtracting the AC gain to the common-mode gain, it is possible to estimate a CMRR of more than 100 dB at 1 Hz. In particular, it is depicted a CMRR of approximately 105 dB, which translates into a considerable robustness to the common-mode voltage, despite the biasing is not in current. The obtained value goes beyond what is presented in [4], i.e., a CMRR of 62.3 dB. Considering a center frequency of noise analysis of approximately 20 Hz, an integrated flicker noise of 11.93 μV_{rms} can be achieved with this topology, which is a particularly important matter, due to the large number of devices connected to the inputs of the amplifier. A thermal noise of approximately 53.87 μV_{rms} is simulated. A summary of state-of-the-art noise contributions in recent literature is presented in Table 5.5, where it is possible to verify that the proposed topology is in line with the prior art, in terms of noise performance. Additionally, a 3-σ Monte Carlo simulation with 100 runs is carried out to acknowledge the robustness of the circuit, and results are shown in Fig. 5.9.

The complete the test, PCB follows the previous equivalent, as described in the previous section. The current consumption of the OTA cannot be directly measured, because the common-mode voltage generation circuitry has to be enabled for the OTA to be operational. Therefore, the current consumption of the circuit is measured

Table 5.5 Comparison with prior art contributions in terms of noise performance

Work	Center frequency	Bandwidth	Flicker	Thermal	Integrated noise
[5]	–	1 Hz–100 MHz	–	–	48.5 μV_{rms}
[6]	–	1 Hz–100 MHz	–	–	118 μV_{rms}
VC-biased OTA	15 Hz	1 Hz–170 MHz	15.66 μV_{rms}	80.9 μV_{rms}	96.56 μV_{rms}
VC-biased OTA w/ current starving	20 Hz	1 Hz–200 MHz	11.93 μV_{rms}	53.87 μV_{rms}	65.8 μV_{rms}

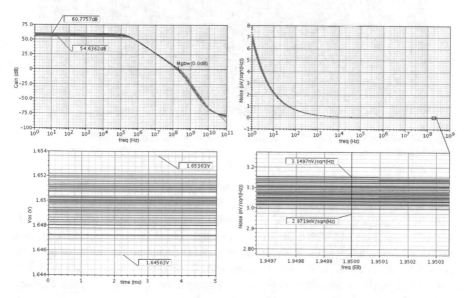

Fig. 5.9 Simulated VC-biased OTA with current starving post-layout process and mismatch

indirectly by subtracting the current consumption of the common-mode voltage circuitry from the overall current consumption of the PCB similar to the previous case, resulting in 1 mA, which includes the complete power dissipation of the internal CMFB circuit. The common-mode voltage of the OTA is set to 1.65 V. In a closed circuit, the OTA sets an output DC voltage of 1.695 V, which means the systematic offset voltage is roughly 45 mV. This can be justified with fabrication deviations and parasitic resistive effects of the PCB. The step response, as shown in Fig. 5.10, is analyzed in a unitary closed circuit, as in [2], following a similar test bench described in the previous section, illustrated in Fig. 5.4. A settling time of 0.08 µs is measured, for a 100 kHz 500 mV input. A rise time of 14.5 ns and a fall time of 15.2 ns are measured, demonstrating an improvement of the OTA, when compared to the previously achieved 21.9 ns and 33.9 ns of rise and fall time correspondingly. This is corroborated with an enhancement of the GBW presented further. The GBW and low-frequency gain of the circuit are measured using a

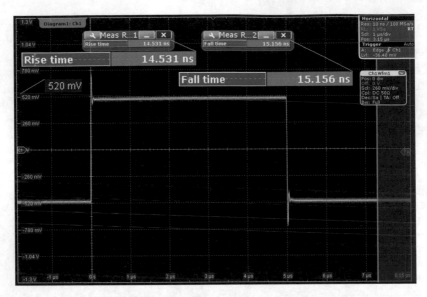

Fig. 5.10 Measured VC-biased OTA with current starving step response: rise time, 14.5 ns; fall time, 15.2 ns

unitary closed-circuit configuration, described in detail in [2] and in [8]. The reference of the network analyzer is applied to the input of the amplifier. The signal is read in the positive output of the setup, with a Hewlett Packard 4195A network/spectrum analyzer and with a 0 dBm input signal. The AC response, with emphasis on the low-frequency gain, is shown in Fig. 5.11, depicting the value of approximately 58 dB. A GBW of 202 MHz is achieved. Once again, a deviation of the GBW can be assigned to a deviation on the true load capacitance of the complete working test bench.

In order to measure the distortion, a 100 mV and 100 kHz sinusoidal signal is set at the input of the OTA in a closed circuit with a gain of 3 V/V. The results are shown in Fig. 5.12, where it is possible to verify that the third harmonic is 50 dBm better than the fundamental. The input swing of the OTA is measured by applying a sweep of sinusoidal signals of 100 kHz in a unitary closed circuit, from 100 mV to 2 V of amplitude. It is verified that above 950 mV, the difference between the second harmonic and the fundamental component is less than 50 dBm; thus, the input swing of the OTA is approximately 950 mV, as in the OTA presented in the previous section. Therefore, no implications on the input swing are verified. The measured noise voltage spectral density is shown in Fig. 5.13. Two additional samples, other than the evaluated in the last paragraphs, have been experimentally measured, and the results are presented in Table 5.6. The variations observed follow the Monte Carlo estimations, depicting a relative deviation of 3.4% and 1.5% in the low-frequency gain and GBW correspondingly.

In order to confirm the potential of both the proposed amplifier topology and the optimization framework that is used in the design process of the circuit, AIDA-C, a

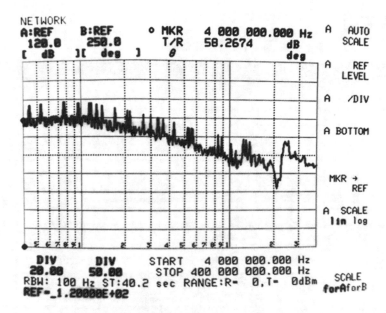

Fig. 5.11   Measured VC-biased OTA with current starving AC gain response

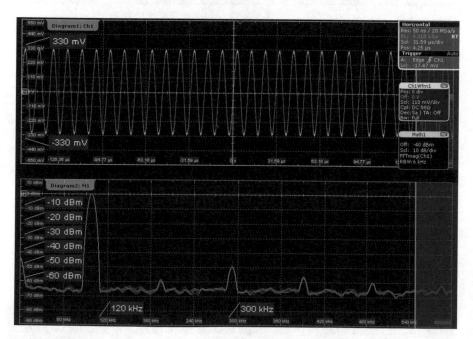

Fig. 5.12   Measured VC-biased OTA with current starving harmonic distortion analysis

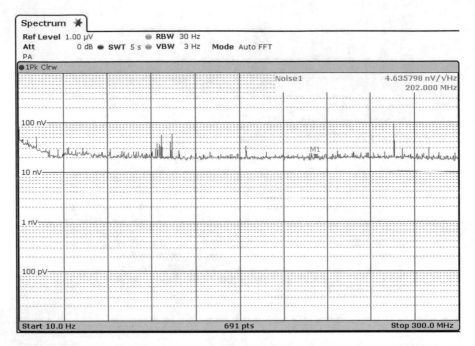

Fig. 5.13 Measured VC-biased OTA with current starving noise voltage spectral density

Table 5.6 Comparison of experimentally measured VC-biased OTA with current starving prototypes

Samples	Gain [dB]	I_{DD} [mA]	PM [°]	GBW [MHz]	FOM [MHz × pF/ mA]	Load [pF]	Tech [nm]
A	58	1.1	>50	202	1102	6	130
B	59	1.1	>50	200	1091	6	130
C	60	1.1	>50	199	1085	6	130

comparison between the fabricated solution and a set of other recently published single-stage amplifiers is presented in Table 5.7. It is possible to verify that this circuit is a relevant contribution to the state of the art in this field. This amplifier is the third most energy-efficient solution, according to (2.14). When compared to the RFCA as presented in [5], the proposed solution has a considerably higher FOM. The purpose of the approach described in this section is demonstrated through Table 5.8. In fact, the GBW of the original OTA is enhanced by a factor of 17.6% and the gain improved in more than 10 dB. Concluding, the energy efficiency improved by approximately 20%.

Table 5.7 VC-biased OTA with current starving compared with recently published single-stage amplifiers

References	Gain [dB]	I_{DD} [mA]	PM [°]	GBW [MHz]	FOM [MHz × pF/ mA]	Load [pF]	Tech [nm]
[5]	60.9	0.800	71	134.2	939.4	5.6	180
[9]	54	1.000	62	965.0	965.0	1.0	180
[6]	85.6	1.000	67	987.0	987.0	1.0	180
[10]	72.0	0.652	70	159.0	536.5	2.2	180
[4]	84.0	0.085	81	12.5	146.0	1.0	180
[11]	94.9	3.670	82	414.0	225.0	2.0	130
[12]	91.5	5.000	62	714.5	1071.5	7.5	130
[13]	67.0	2.170	67	920.0	211.0	0.5	180
VC-biased OTA w/ current starving[a]	57.9	1.270	51.9	195.00	921.3	6	130
VC-biased OTA w/ current starving[b]	58	1.1	>55	202	1102	6	130

[a]Post-layout simulation results
[b]Experimental measurements

5.4 Summary

This chapter presents the practical realization and the experimental measurements of a new family of single-stage OTAs, biased by voltage-combiners, with high gain and energy efficiency. In this new family, the traditional current source that typically biases the differential pair of a symmetrical CMOS OTA is replaced by several configurations of voltage-combiners, i.e., in a cross-coupled configuration, in a folded configuration to tackle specifically low-voltage supplies, and finally in a dynamic configuration to be implemented in high-performance ADCs and SC filters. A summary of the works proposed and respective implementations is presented in Table 5.8, which comprises results at simulation level after optimization and measurement results of the fabricated prototypes.

The potential of the new family of CMOS single-stage amplifiers, developed in this work, is clearly demonstrated and validated among the context of the state of the art of single-stage amplifiers. Indeed, through Table 5.8, it is possible to verify that the proposed topologies and implementations achieve the highest values of energy efficiency among the state of the art while demonstrating competitive values of gain also.

The most important advantages of the proposed topologies, e.g., a high FOM, with a better OS when compared to the limitations of classic cascode architectures, are compounded by the fact that the spent area on-chip is relatively small, i.e., lower than 0.04 mm². Moreover, the intrinsic simplicity of the topologies makes them attractive and forecasts a straightforward implementation in deeper technology

Table 5.8 Comparison of state-of-the-art single-stage amplifiers

References	Gain [dB]	I_{DD} [mA]	PM [°]	GBW [MHz]	FOM [MHz × pF/ mA]	Load [pF]	Tech [nm]	Available OS
[5][a]	60.9[b]/ 53.6[a]	0.800	71	134.2	939.4	5.6	180	$OS \approx V_{DD} - 4 \times VDS_{SAT}$
[9][a]	54	1.000	62	965.0	965.0	1.0	180	$OS \approx V_{DD} - 4 \times VDS_{SAT}$
[6][b]	85.6	1.000	67	987.0	987.0	1.0	180	$OS \approx V_{DD} - 4 \times VDS_{SAT}$
[10][b]	72.0	0.652	70	159.0	536.5	2.2	180	$OS \approx V_{DD} - 4 \times VDS_{SAT}$
[4][a]	84.0	0.085	81	12.5	146.0	1.0	180	$OS \approx V_{DD} - 4 \times VDS_{SAT}$
[11][b]	94.9	3.670	82	414.0	225.0	2.0	130	$OS \approx V_{DD} - 4 \times VDS_{SAT}$
[12][b]	91.5	5.000	62	714.5	1071.5	7.5	130	$OS \approx V_{DD} - 4 \times VDS_{SAT}$
[13][b]	67.0	2.170	67	920.0	211.0	0.5	180	$OS \approx V_{DD} - 2 \times VDS_{SAT}$
VC-biased OTA[a]	47	1	>55	170.6	1023.6	6	130	$OS \approx V_{DD} - 2 \times VDS_{SAT}$
VC-biased OTA w/current starving[a]	58	1.1	>55	202	1102	6	130	$OS \approx V_{DD} - 2 \times VDS_{SAT}$
Folded VC-biased OTA[b]	54.7	0.215	63.5	51.32	1435.3	6	130	$OS \approx V_{DD} - 2 \times VDS_{SAT}$
Folded VC-biased OTA[b]	50.3	0.218	61	82.79	2278.5	6	130	$OS \approx V_{DD} - 2 \times VDS_{SAT}$
[14][b]	95	23	59	1480	128.7	2	90	$OS \approx V_{DD} - 2 \times VDS_{SAT}$
Dynamic VC-biased OTA SE[b]	61	0.235	44.8	51.73	1320.8	6	130	$OS \approx V_{DD} - 2 \times VDS_{SAT}$
Dynamic VC-biased OTA FD[b]	60.9	0.69	44	155.1	1349	6	130	$OS \approx V_{DD} - 2 \times VDS_{SAT}$

[a]Experimental measurements
[b]Simulation results

nodes and lower supply voltages. These aspects are, therefore, in line and according to the progressively higher demand and need to embed more circuits in smaller chips, with emphasis on the energy efficiency of the circuits in the wide context of the IOT. Specifically, regarding the dynamic VC-biased OTA, the values of FOM are the state of the art in the field of dynamic OTA implementations. When compared to the work proposed in [14], which is a two-stage OTA, which justifies the gain of 95 dB, the proposed solution has a much higher energy efficiency, which derives greatly from the low-power consumption, even considering the technology node. Moreover, the input-referred noise of the proposed solution represents approximately 40% of the value shown in [14], with an achieved gain above 60 dB.

References

1. Online, 2017: https://www.autodesk.com/products/eagle/overview.
2. Online, 2017: https://www.ti.com/lit/ds/symlink/ths4524.pdf.
3. R. Baker, "CMOS Circuit Design, Layout and Simulation," Third Edition, Wiley, 2010. ISBN: 978-0-470-88132-3.
4. M. Fallah, H. Naimi, "A Novel Low Voltage, Low Power and High Gain Operational Amplifier Using Negative Resistance and Self Cascode Transistors," in International Journal of Engineering Transactions C: Aspects, Vol. 26, Issue 3, Page(s): 303–308, 2013.
5. R. Assaad, J. Silva-Martinez, "The Recycling Folded-cascode: A General Enhancement of the Folded-Cascode Amplifier," in IEEE Journal of Solid-State Circuits, Vol. 44, Issue 9, Page(s): 2535–2542, Sep. 2009. DOI: https://doi.org/10.1109/JSSC.2009.2024819.
6. M. Ahmed, et al., "An Improved Recycling Folded-cascode Amplifier with Gain-boosting and Phase Margin Enhancement," in IEEE International Symposium on Circuits and Systems (ISCAS), Page(s): 2473–2476, May 2015. DOI: https://doi.org/10.1109/ISCAS.2015.7169186.
7. Online, 2017: http://www.analog.com/media/en/training-seminars/tutorials/MT-042.pdf.
8. M. Figueiredo, et al., "A Two-Stage Fully Differential Inverter-Based Self-Biased CMOS Amplifier With High Efficiency," in IEEE Transactions on Circuits and Systems-I: Regular Papers, Vol. 58, Issue 7, Page(s): 1591–1603, July, 2011. DOI: https://doi.org/10.1109/TCSI.2011.2150910.
9. Y. Li, et al., "Transconductance enhancement method for operational transconductance amplifiers," in Electronics Letters, Vol. 46, Issue 19, Page(s): 1321–1323, Sep. 2010. DOI: https://doi.org/10.1049/el.2010.1575.
10. S. Zabihian, R. Lofti, "Ultra-Low-Voltage, Low-Power, High-Speed Operational Amplifiers Using Body-Driven Gain-Boosting Technique," in IEEE International Symposium on Circuits and Systems (ISCAS), Page(s): 705–708, May 2007. DOI: https://doi.org/10.1109/ISCAS.2007.377906.
11. S. Enche, et al., "A CMOS Single-Stage Fully Differential Folded-cascode Amplifier Employing Gain-boosting Technique," in IEEE International Symposium on Integrated Circuits (ISIC), Page(s): 234–237, Dec. 2011. DOI: https://doi.org/10.1109/ISICir.2011.6131939.
12. X. Liu, J. McDonald, "Design of Single-Stage Folded-Cascode Gain-boost Amplifier for 14bit 12.5Ms/S Pipelined Analog-to-Digital Converter," in IEEE International Conference on Software Engineering, Page(s): 622–626, Sep. 2012. DOI: https://doi.org/10.1109/SMElec.2012.6417222.

13. B. Alizadeh, A. Dadashi, "An Enhanced Folded-cascode Op Amp in 0.18 μm CMOS Process with 67dB Dc Gain," in IEEE Faible Tension Faible Consommation (FTFC), Page(s): 87–90, May 2011. DOI: https://doi.org/10.1109/FTFC.2011.5948926.
14. A. Zadeh, "A 100MHz, 1.2V, ±1V Peak-to-Peak Output, Double-Bus Single Ended-to-Differential Switched Capacitor Amplifier for Multi-Column CMOS Image Sensors," in IEEE International New Circuits and Systems Conference (NEWCAS), Page(s): 1–4, Jun. 2016. DOI: https://doi.org/10.1109/NEWCAS.2016.7604739.

Chapter 6
Conclusions and Future Prospects

In recent years, electronics designers, especially designers of amplifiers, have been struggling to overcome the challenges associated with the reduction of the intrinsic gain of active devices, i.e., transistors, with even higher impact on deeper nanoscale technologies. On another hand, the nowadays technological market demands circuits and systems with elevated energy efficiency and possibly small area, in order to ensure portability and comfort for the daily users. Particularly, in the context of mixed signal, i.e., ADC architectures, designers have been seeking for alternative ways to avoid the usage of amplifiers, in order to improve the power consumption, increasing the complexity of the circuits, and hence lowering their reliability with also a direct penalty in terms of area spent. This work contributes to the advancement of the state of the art, by proposing a complete new family of single-stage transconductance amplifiers with elevated energy efficiency, with also an innovative gain enhancement strategy that avoids the need for cascode devices, i.e., the output signal swing is improved and, therefore, is clearly more appropriate to follow the tendency of using low voltage supplies, in nowadays circuitry. Both simulation and experimental evaluation are presented.

In Chap. 2, a detailed explanation of the most important metrics to take into account in the context of amplifiers was first presented. Afterward, a comprehensive study and analysis of the state of the art, in the vast field of single-stage amplifiers, was offered. Moreover, in this chapter, a complete description of amplifier architectures, comprising single-ended and fully-differential single-stage amplifiers, was presented, with emphasis in the cascode topologies and in the dynamic biasing concept, which relevance is restored in the present time, in the context of low-power circuits and systems. In Chap. 3, the innovative techniques proposed in this work were presented and analyzed. In particular, the usage of voltage-combiners, in replacement of the traditional static current sources that bias the differential pairs of classic single-stage amplifiers, was proposed and compounded with preliminary simulation results. The proposed change has a twofold effect: (a) additional gain is provided; (b) the differential-pair devices act as a common source and a common

© Springer International Publishing AG, part of Springer Nature 2019
R. F. S. Póvoa et al., *A New Family of CMOS Cascode-Free Amplifiers with High Energy-Efficiency and Improved Gain*,
https://doi.org/10.1007/978-3-319-95207-9_6

gate simultaneously; this way the energy efficiency of the circuit is improved by means of increasing the gain-bandwidth product. Also in this chapter, the topologies proposed in this work were presented in detail, analyzed, and validated at simulation level: the voltage-combiner-biased OTA; the voltage-combiner-biased OTA with current starving, which corresponds to an upgrade of the latter when higher gains are required and a higher area is available; the folded voltage-combiner-biased OTA, which targets low-voltage applications; and finally the dynamic voltage-combiner-biased OTA, which addresses high-speed and low-power SC circuits. For proof of concept, the UMC 130 nm CMOS technology node was used in this work. The migration to deeper nanoscale nodes is relatively straightforward. In Chap. 4, the automatic IC design framework that was used to optimize the performance of the proposed topologies was described. AIDA is a multi-objective and multi-constraint circuit optimization framework that automatically synthesizes electronic circuits at sizing level and also at layout level. However, regarding this work, only the sizing optimization was considered, i.e., all the layouts were designed manually. Detailed setup definitions, optimization strategies, and the results after optimization were presented for all topologies. In the case of the folded voltage-combiner-biased OTA and in the case of the dynamic voltage-combiner-biased OTA, gains above 50 dB and 60 dB, respectively, and FOMs of 2279 MHz × pF/mA and 1349 MHz × pF/mArms were achieved, likewise, which are clear contributions to the state of the art in the field of single-stage amplifiers, both in terms of static and dynamic biasing. In Chap. 5, the integrated prototypes were described in detail, comprising the layout design and post-layout simulations. The design of the PCB dedicated to the experimental evaluation of the integrated prototypes was described, and the experimental evaluation of the prototypes was addressed, resulting in measured gains of 47 and 58 dB and FOMs of 1024 MHz × pF/mA and 1102 MHz × pF/mA, for the voltage-combiner-biased OTA and voltage-combiner-biased OTA with current starving, correspondingly, which are also clear contributions to the state of the art in the field of single-stage amplifiers, namely, in terms of energy efficiency.

In summary, two topologies were proposed and validated at simulation level, i.e., the folded voltage-combiner-biased OTA and the dynamic voltage-combiner-biased OTA, both biased with 1.2 V voltage supplies; and two topologies were proposed and validated with experimental results, i.e., the voltage-combiner-biased OTA and the voltage-combiner = biased OTA with current starving, also with elevated energy-efficiency, both biased with 3.3 V nominal voltage power supplies. In a final note, two integrated prototypes have been tested with relevant results when compared to the state of the art, and a specific topology properly developed for low-voltage implementations has been designed and explored with encouraging results at simulation level. Moreover, a dynamic single-stage amplifier, targeting ADCs and SC filters, has been implemented, with state-of-the-art results at simulation level. In conclusion, an innovative gain enhancement technique, through the replacement of a static current source by voltage-combiners, has been demonstrated with success, with clear benefits in the context of single-stage amplifiers.

With the proposed topologies, higher energy-efficient values can be achieved, together with higher output swing, especially when compared to cascode-based topologies, looking toward the broad usage of lower-voltage supplies in the near future of electronics design.

Biographies

Ricardo Filipe Sereno Póvoa received the BSc, MSc, and PhD degrees in Electrical and Computer Engineering from Instituto Superior Técnico (IST), Universidade de Lisboa, Portugal, in 2011, 2013, and 2018, respectively. Since 2013, he has a research position at the Instituto de Telecomunicações, where he is now a postdoctoral researcher, working on analog and mixed-signal circuit design. In February 2016, he joined the Marine Engineering Department of Escola Superior Náutica Infante D. Henrique, where he teaches Naval Electronics and Automation courses. At Instituto de Telecomunicações, he is currently working in the field of low-power IC design, applied to biometric applications and healthcare. His research interests are mainly in analog and mixed-signal circuit design, RF CMOS circuitry and applications, and electronic design automation.

João Carlos da Palma Goes João Carlos da Palma Goes (S'95–M'00–SM'09) graduated from Instituto Superior Técnico (IST), Universidade de Lisboa, Portugal, in 1992, and received the MSc and PhD degrees from the same university in 1996 and 2000, respectively. From December 1993 to February 1997, he worked as a Senior Researcher at Integrated Circuits and Systems Group (ICSG) at IST doing research on data converters and analog filters. From March 1997 until March 1998, he was the Project Manager at Chipidea SA. He has been with the Department of Electrical Engineering (DEE) of the Faculdade de Ciências e Tecnologia (FCT) of Universidade Nova de Lisboa (UNL) since April 1998, where he is currently an Associate Professor. Since 1998 he has been a Senior Researcher at the Center for Technology and Systems (CTS) at UNINOVA. In 2003 he cofounded and served as the CTO of ACACIA Semiconductor, a Portuguese engineering company specialized in high-performance data converter and analog front-end products (acquired by Silicon and Software Systems, S3, in October 2007). Since November 2007 he does his lectures with part-time consultancy work for S3. Since 1992 he has participated and led several national and European projects in science, technology, and training. His scientific interests are in the areas of low-power and low-voltage analog

© Springer International Publishing AG, part of Springer Nature 2019
R. F. S. Póvoa et al., *A New Family of CMOS Cascode-Free Amplifiers with High Energy-Efficiency and Improved Gain*,
https://doi.org/10.1007/978-3-319-95207-9

integrated circuits, data converters, built-in self-test and self-calibration techniques, and optimization and automatic sizing of analog circuits. He has published over 70 papers in international journals and leading conferences, and he is a co-author of *Systematic Design for Optimization of Pipelined ADCs* (Springer, 2001), *Low Power UWB CMOS Radar Sensors* (Springer, 2008), and *VLSI Circuits for Biomedical Applications* (Artech House, 2008). Prof. Goes has been a member of the Portuguese Professional Association of Engineers since 1992. He has served as a reviewer for journals and conferences, and he is a member of the Technical Program Committee of ISCAS (IEEE) and AMICSA (ESA) Conferences.

Nuno Cavaco Gomes Horta $(S'89-M'97-SM'11)$ received the Licenciado, MSc, PhD, and Habilitation degrees in Electrical Engineering from Instituto Superior Técnico (IST), Universidade de Lisboa, Portugal, in 1989, 1992, 1997, and 2014, respectively. In March 1998, he joined the IST Electrical and Computer Engineering Department. Since 1998, he is, also, with Instituto de Telecomunicações, where he is the head of the Integrated Circuits Group. He has supervised more than 90 postgraduation works between MSc and PhD theses. He has authored or co-authored more than 150 publications such as books, book chapters, international journal papers, and conference papers. He has also participated as researcher or coordinator in several national and European R&D projects. He was the General Chair of AACD 2014, PRIME 2016, and SMACD 2016 and was a member of the organizing and technical program committees of several other conferences, e.g., IEEE ISCAS, IEEE LASCAS, DATE, NGCAS, etc. He is the Associate Editor of Integration, The VLSI Journal, from Elsevier, and usually acts as reviewer of several prestigious publications, e.g., IEEE TCAD, IEEE TEC, IEEE TCAS, ESWA, ASC, etc. His research interests are mainly in analog and mixed-signal IC design, analog IC design automation, soft computing, and data science.

Index

A
AC, 14, 25, 49, 56, 57, 59, 60, 64, 65, 68, 69, 71, 99, 100, 117, 119, 121, 124, 126, 127
AIDA, 4, 85–88, 90, 91, 93, 95, 111, 126, 134
Amplifier, 4, 7–11, 13–21, 23–25, 27–32, 34–39, 41, 45, 47, 52, 53, 57–59, 61, 62, 64, 66–68, 73, 78, 79, 90, 91, 98, 115–117, 119, 122–124, 126, 133, 134
Automation, 2, 90

B
Body effect, 11, 12, 38, 45–47, 53, 55, 59, 61, 64, 67, 68, 80, 88, 91, 93, 98
Burst noise, 74, 75

C
Capacitors, 13, 34, 67, 68
Cascode, 1, 4, 7, 8, 24–34, 36, 38, 42, 52, 53, 58, 62, 66, 78, 129, 133, 135
Channel length, 64, 65
Channel width, 76
CMFB, 55, 56, 59, 89, 91, 93, 95, 99, 119, 125
CMOS, 1, 4, 8, 16, 18, 22–24, 29, 31, 39–41, 45, 47, 53, 57, 58, 65, 74, 76–79, 83, 87, 88, 90, 93, 111, 129, 134
CMRR, 16, 17, 73, 118, 124
CNT, 104
Common-drain (CD), 12, 45–47, 52, 61, 62, 66
Common-gate (CG), 4, 12, 25, 26, 38, 50, 52, 53, 61, 66, 133

Common-source (CS), 4, 9–11, 13, 14, 23, 25, 30, 36, 38, 45–47, 50, 52, 53, 61, 66, 77, 78, 133
Cross-coupled, 35, 52, 53, 56, 57, 67, 73, 83, 111, 129
Current, 4, 8–11, 13, 15, 16, 18, 21–23, 25–27, 29, 31, 34–36, 38–41, 45, 47, 52–55, 57–62, 64–68, 70, 72, 73, 75, 76, 79, 81, 83, 86, 88, 90–92, 94, 99, 100, 111, 115, 117, 119, 122–129, 133, 134
Current starving, 58, 61, 116, 122, 134

D
DC, 9, 10, 13, 16, 19, 22, 27, 47, 52, 58, 64, 69, 72, 75, 87–89, 91, 99, 100, 119, 125
Differential-pair, 4, 13, 23, 28, 34, 52, 53, 56–58, 61, 62, 64, 66, 67, 69, 81, 83, 99, 111, 129, 133
Dynamic, 4, 39–41, 47, 63, 65–67, 69–73, 83, 95, 99–102, 111, 129, 131, 133, 134

E
Electronic design automation, 101, 103, 137
Energy-efficiency, 1, 3, 4, 7, 15, 23, 38, 41, 47, 52, 57, 66, 67, 83, 85, 88, 90, 95, 111, 117, 128, 129, 131, 133–135
Experimental, 4, 55, 58, 63, 68, 116, 120, 122, 129, 133, 134

© Springer International Publishing AG, part of Springer Nature 2019
R. F. S. Póvoa et al., *A New Family of CMOS Cascode-Free Amplifiers with High Energy-Efficiency and Improved Gain*,
https://doi.org/10.1007/978-3-319-95207-9

Printed in the United States
By Bookmasters